Electrical Networks: Theory and Analysis

Electrical Networks: Theory and Analysis

Britney Headley

Willford Press
www.willfordpress.com

Published by Willford Press,
118-35 Queens Blvd., Suite 400,
Forest Hills, NY 11375, USA

ISBN: 978-1-64728-507-4

Cataloging-in-Publication Data

Electrical networks : theory and analysis / Britney Headley.
p. cm.
Includes bibliographical references and index.
ISBN 978-1-64728-507-4
1. Electric networks. 2. Electric network analysis. 3. Electric power distribution.
4. Electric lines. 5. Electric circuits. I. Headley, Britney.
TK454.2 .E44 2023
621.319 2--dc23

For information on all Willford Press publications
visit our website at www.willfordpress.com

Willford Press

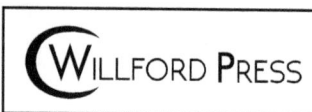

Contents

Preface

An electrical network, also called electric circuit or electrical circuit, refers to an interconnection of various active and passive components in a prescribed manner to form a closed path. The main types of electric circuits are open circuits, closed circuits, short circuits, series circuits, parallel circuits, and series parallel circuits. Network theory deals with the study of solving the problems related to electric networks or electronic networks. The process through which different electrical parameters of a circuit element connected in an electrical network are calculated is called network analysis. The circuit components such as resistors, capacitors, inductors, voltage sources, current sources can either be connected in series or in parallel combinations. Network analysis determines the different electrical parameters including current, voltage, resistance, impedance, reactance, inductance, capacitance, frequency, electric power and electrical energy. This book is a compilation of chapters that discuss the theory and analysis of electrical networks. It aims to serve as a resource guide for scholars and engineers in the fields of electronics and communication engineering as well as electrical engineering.

After months of intensive research and writing, this book is the end result of all who devoted their time and efforts in the initiation and progress of this book. It will surely be a source of reference in enhancing the required knowledge of the new developments in the area. During the course of developing this book, certain measures such as accuracy, authenticity and research focused analytical studies were given preference in order to produce a comprehensive book in the area of study.

This book would not have been possible without the efforts of the authors and the publisher. I extend my sincere thanks to them. Secondly, I express my gratitude to my family and well-wishers. And most importantly, I thank my students for constantly expressing their willingness and curiosity in enhancing their knowledge in the field, which encourages me to take up further research projects for the advancement of the area.

Britney Headley

1

Electrical Circuits, AC Fundamentals and Network Topology

1.1 Electrical Circuits

An electric circuit or electric network is the interconnection of electrical elements linked together in the closed path so that an electric current may continuously flow. Also, an electric circuit is essentially a pipe-line which facilitates the transfer of charge from one point to another.

1.1.1 Network Elements Classification

The most elementary quantity in the analysis of the electric circuits is the electric charge. Charge is the intrinsic property of the matter responsible for electrical phenomena. The quantity of charge q can be expressed in terms of charge on one electron which is -1.602×10^{-19} coulombs.

Thus, -1 coulomb is the charge on 6.24×10^{18} electrons. The current flows through the specified area A and it can be defined by the electric charge passing through that area per unit time. Thus, we can define q as the charge expressed in coulombs.

1.1.2 Electric Charge and Current

Electric Charge is the quantity of electricity responsible for electric phenomena. The time rate of change constitutes an electric current. Mathematically, this relation can be expressed as,

$$i(t)=\frac{dq(t)}{dt}$$

$$q(t)=\int_{-\infty}^{t} i(x)dx$$

$i(t)$ = Current flowing in the circuit

$dq(t)$ = Amount of charge passing through the circuit

dt = Time taken by the charge

The unit of current is expressed as ampere (A). An ampere is 1 coulomb per second.

Current is the time rate of flow of electric charge past a given point. If the current flows into the terminal of the element shown in the below figure, then a voltage or potential difference exists between the two terminals a and b. normally, we say that a voltage exists across the element.

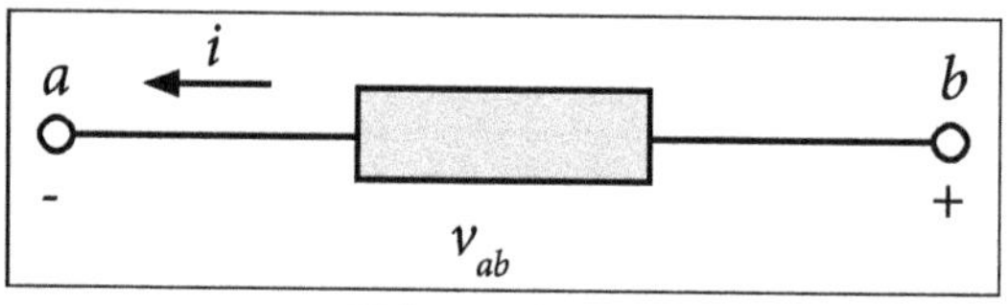

Voltage circuit.

The voltage across an element is the work done in moving a positive charge of 1 coulomb from the first terminal through the element to the second terminal. The unit of voltage is expressed in volt, V or Joules per coulomb.

We have defined the voltage in Joules per coulomb as the energy required to move a positive charge of 1 coulomb through the element. If we assume that we are dealing with the differential amount of charge and energy, then,

$$v = \frac{dw}{dq}$$

By multiplying both the sides of the above equation by current in the element,

We get,

$$vi = \frac{dw}{dq}\left(\frac{dq}{dt}\right) \Rightarrow \frac{dw}{dt} = p$$

p could be either positive or negative. Thus, it is imperative to give sign convention for power.

If we use the signs as shown in the below figure, the current flows out of the terminal indicated by x, which shows the positive sign for the voltage. In this case, the element provides energy to the charge as it moves through. Power is then provided by the element.

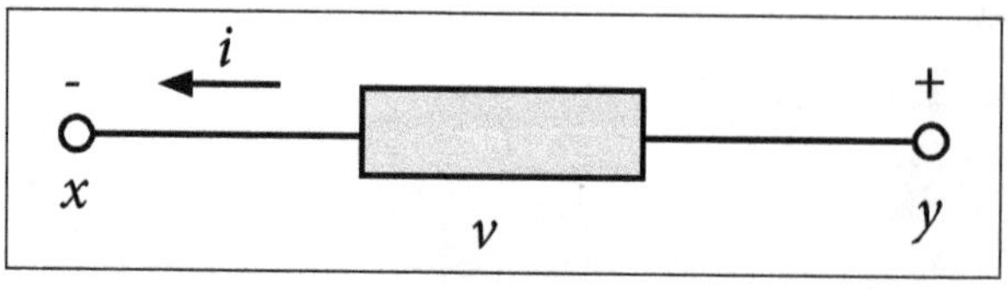

An element with the current leaving from the terminal with a positive voltage sign.

Conversely, the power absorbed by an element is expressed as p = vi, when i is entering through the positive voltage terminal.

1.1.3 Electric Energy and Potential

Energy is the capacity to perform work. Energy and power are related to each other by the following equation,

$$\text{Energy} = w = \int_{-\infty}^{t} p(dt)$$

Electric Potential

Electric potential is more commonly known as voltage. The potential at a point a distance r from a charge Q is given by,

$$V = k\,Q/r$$

Where,

$$K = \frac{1}{4\pi\varepsilon_0}$$

$$\varepsilon_0 = \text{Permittivity of vacuum}$$

Potential plays the same role for charge that pressure does for fluids. If there is a pressure difference between two ends of a pipe filled with fluid, the fluid will flow from the high pressure end towards the lower pressure end. Charges respond to differences in potential in a similar way.

Electric potential is a measure of the potential energy per unit charge. If the potential at a point is known, when placing a charge at that known point, the potential energy associated with that charge in that potential is simply the charge multiplied by the potential. Electric potential, like potential energy, is a scalar, not a vector.

Connection between potential and potential energy is given as,

$$V = PE/q$$

Equipotential lines are the connected lines of same potential. These often appear on field line diagrams. Equipotential lines are always perpendicular to field lines and therefore perpendicular to the force experienced by a charge in the field. If a charge moves along an equipotential line, no work is done, if a charge moves between equipotential lines, work is done.

Field lines and equipotential lines for a point charge and for a constant field between two charged plates, as shown below,

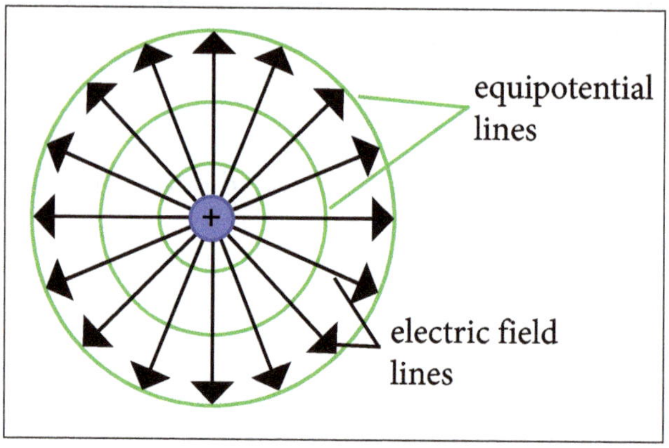

Field and equipotential lines for a positive point charge.

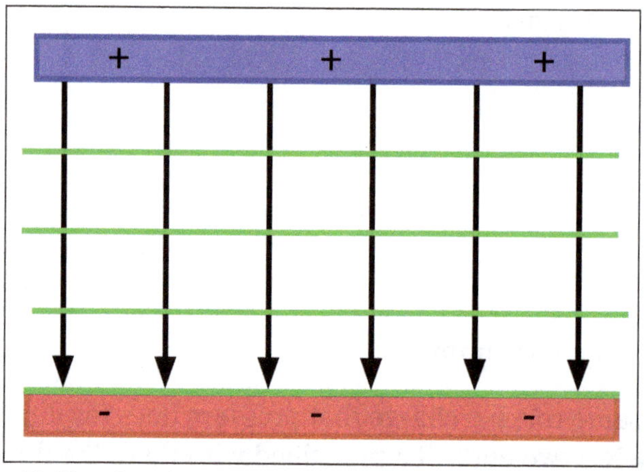

Field and equipotential lines for a set of parallel plates.

An Example: Ionization Energy of the Electron in a Hydrogen Atom

In the Bohr model of a hydrogen atom, if the electron is in the ground state, orbits the proton at a distance of $r = 5.29 \times 10^{-11}$ m. It is important to note that in the Bohr model, the idea of electrons as tiny balls orbiting the nucleus is not a very good model of the atom.

A better picture is one in which the electron is spread out around the nucleus in a cloud of varying density. However, the Bohr model gives the right answer for the ionization energy, the energy required to remove the electron from the atom.

The total energy is the sum of the electron's kinetic energy and the potential energy coming from the electron-proton interaction.

The kinetic energy is given by $KE = 1/2\ mv^2$.

This can be found by analyzing the force on the electron. This force is the Coulomb force, because the electron travels in a circular orbit, the acceleration will be the centripetal acceleration,

$$\Sigma F = ma$$

So,

$$\Sigma F = k\,e^2/r^2 = ma = mv^2/r$$

It is clear that the negative sign coming from the charge on the electron has been incorporated into the direction of the force in the equation above.

This gives $mv^2 = k\,e^2 / r$, so the kinetic energy is $KE = 1/2\,k\,e^2 / r$.

The potential energy, on the other hand, is $PE = -\,k\,e^2/r$. It is important to note that the potential energy is twice as big as the kinetic energy, but negative. This relationship between the kinetic and potential energies is valid not just for electrons orbiting protons, but also in gravitational situations, such as satellite orbiting the Earth.

The total energy is given by,

$$KE + PE = -1/2\;ke^2/r = -1/2\;(8.99 \times 10^9)(1.60 \times 10^{-19})/5.29{\times}10^{-11}$$

This works out to -2.18×10^{-18} J. This is usually stated in energy units of electron volts (eV). An eV is 1.60×10^{-19} J, so dividing by this gives an energy of -13.6 eV. To remove the electron from the atom, 13.6 eV must be used, 13.6 eV is thus the ionization energy of a ground-state electron in hydrogen.

1.2 Resistance Parameter, Inductance Parameter and Capacitance Parameter: Series and Parallel Combination

Basic Circuit Components:

- Resistor
- Capacitor
- Inductor

1.2.1 Resistance Parameter: Series and Parallel Combination

Resistor

It is defined as the property of the material by which it opposes the flow of current through it:

- It is denoted by R.

- It's unit is measured in (Ω) Ohms.

Symbol

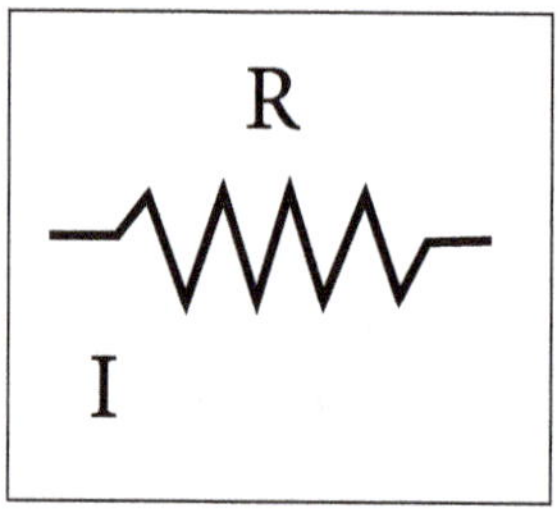

Resistor

The relation between V & I is given by Ohm's law,

$$V = IR$$

Resistance in Series

When resistors are connected in series, same current passes through all of them.

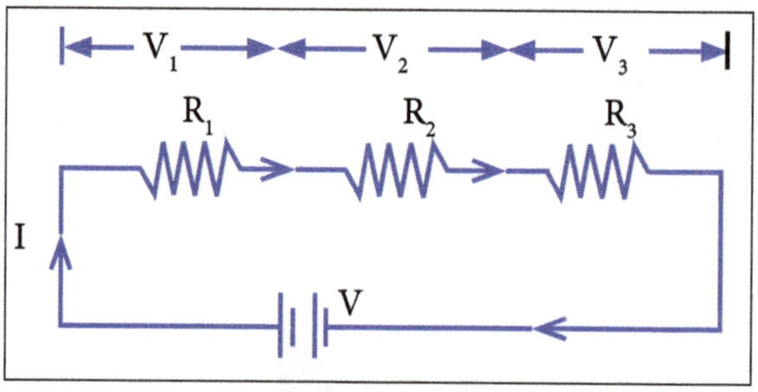

Resistance in series.

Where,

$R_1, R_2, R_3 \rightarrow$ Resistor in Series
$V_1, V_2, V_3 \rightarrow$ Voltage
$I \rightarrow$ Current

Then,

$$V = V_1 + V_2 + V_3$$

By Ohm's law,

$$V_1 = IR_1;\ V_2 = IR_2;\ V_3 = IR_3\,[V = IR]$$

$$V = IR_1 + IR_2 + IR_3$$

$= I\,(R_1 + R_2 + R_3)$

$= I.\, R_{eq} \rightarrow$ equivalent resistor of the current

$R_{eq} = R_1 + R_2 + R_3$

Resistance in Parallel

When resistor are connected in parallel, same voltage appears across each of them.

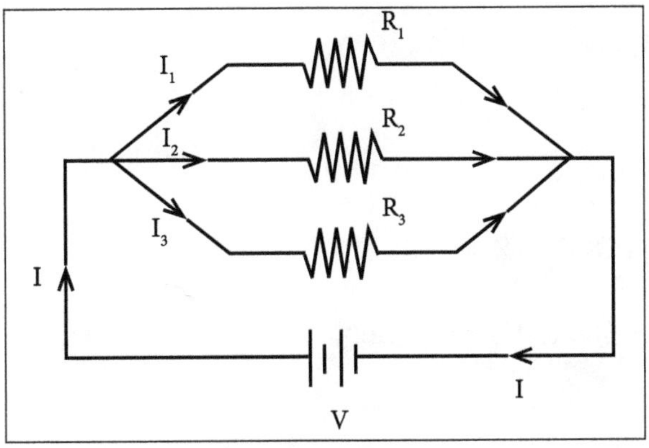

Resistance in parallel.

$R_1, R_2, R_3 \rightarrow$ Three resistances in parallel

$I_1, I_2, I_3 \rightarrow$ Individual current

$V \rightarrow$ Voltage

Then,

$I \rightarrow I_1 + I_2 + I_3$

By ohm's law,

$I_1 = V/R_1$

$I_2 = V/R_2$

$I_3 = V/R_3$

$I = V/R_1 + V/R_2 + V/R_3$

$= V\left[1/R_1 + 1/R_2 + 1/R_3\right]$

$$= V.\left[1/R_{eq}\right]$$

$I = V/R^{eq} \rightarrow$ Equivalent resistance of the parallel

$$I/R_{eq} = 1/R_1 + 1/R_2 + 1/R_3$$

If 2 resistor are connected in parallel, then,

$$I/R_{eq} = 1/R_1 + 1/R_2$$
$$I/R_{eq} = R_1 + R_2 / R_1R_2$$

If 3 resistor are connected in parallel, then,

$$I/R_{eq} = 1/R_1 + 1/R_2 + 1/R_3$$

$$1/R_{eq} = R_2R_3 + R_1R_3 + R_1R_2 / R_1R_2R_3$$

$$R_{eq} = R_1R_2R_3 / R_2R_3 + R_1R_2 + R_3R_1$$

If 2 resistors are equal (i.e.,), $R_1 = R_2$ then,

$$R_{eq} = R_1 / 2.$$

1.2.2 Inductance Parameter: Series and Parallel Combination

The inductor is a coil which stores the energy in the magnetic field. Let us consider a wire of length l forming a loop of area A as shown in the below figure.

A current I (t) is flowing through the wire as indicated in the below figure. This current generates the magnetic field B which is equal to the given relation,

$$B(t) = \mu \frac{i(t)}{l}$$

Where,

μ is the magnetic permeability of the material enclosed by the wire.

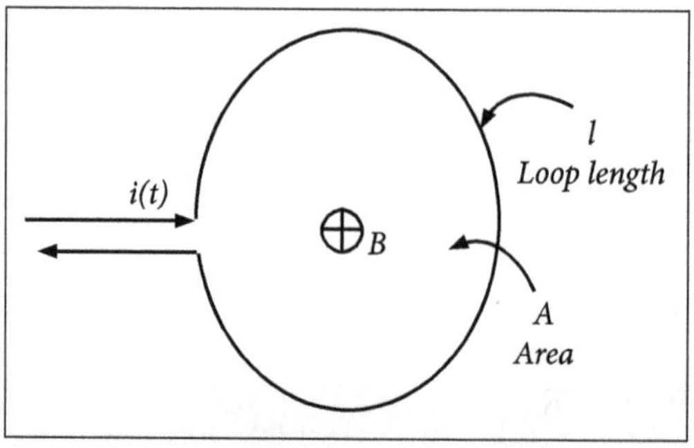

Current loop for the calculation of inductance.

The magnetic flux ϕ, through the loop of the area A is denoted as,

$$\Phi = AB(t)$$

$$= \frac{A\mu}{l} i(t)$$

$$= Li(t)$$

Where we have defined,

$$L = \frac{A\mu}{l}$$

From Maxwell's equations, we know that,

$$\frac{d\Phi}{dt} = v(t)$$

$$\frac{d\,Li(t)}{dt} = v(t)$$

And by taking L as a constant, we can obtain the current-voltage relationship for this loop of wire also known as an inductor,

$$v = L\frac{di}{dt}$$

The parameter L is known as the inductance of the inductor. Its unit is Henry (H). The circuit symbol and associated electrical variables for the inductor is shown in the below figure.

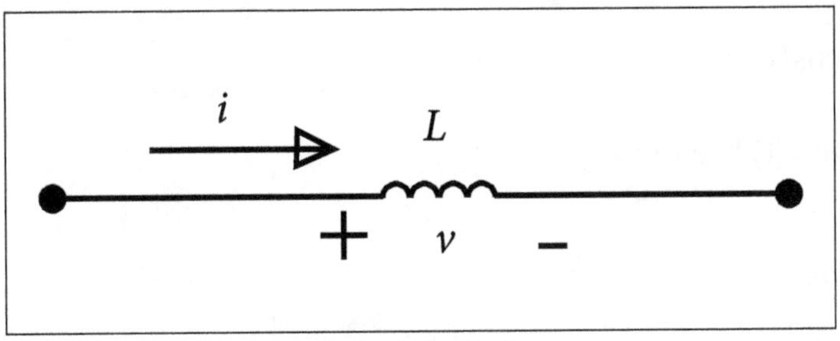

Circuit symbol of inductor.

For DC signals ($di/dt = 0$), the inductor acts as a short circuit ($v = 0$). The inductor does not like current discontinuities, since that would require the voltage across it, which may goes to infinity and it is not physically possible.

If we integrate the above equation over time,

We have,

$$\int_{-\infty}^{t} v\,dt = \int_{-\infty}^{t} L\frac{di}{dt}dt$$

$$i = \frac{1}{L} = \int_{-\infty}^{t} v\,dt$$

$$= \frac{1}{L}\int_{0}^{t} v\,dt + i(0)$$

The constant i(o) represents the current through the inductor at time t = o. We have also assumed that the current at t = −∞ was zero.

Let us now consider the circuit shown in the below figure where an inductor of inductance L is connected to a time varying current source i(t).

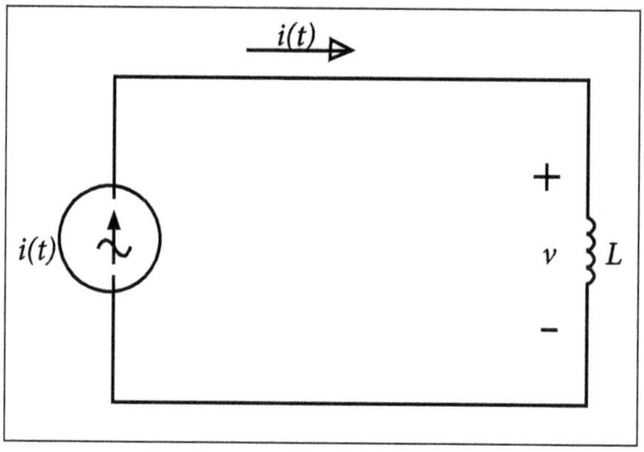

Fundamental inductor circuit.

If we assume that the current i(t) has the form,

$$i(t) = I_0 \cos(\omega t)$$

Then, the voltage v(t) becomes,

$$v(t) = L\frac{di}{dt}$$

$$= -LI_0\,\omega \sin(\omega t)$$

$$= L\omega I_0 \cos\left(\omega t + \frac{\pi}{2}\right)$$

Thus, the current going through an inductor and the voltage across the inductor are 90 degrees out of phase. Here the voltage leads the current by 90 degrees.

The general plot of the voltage and current of an inductor is shown in the below figure.

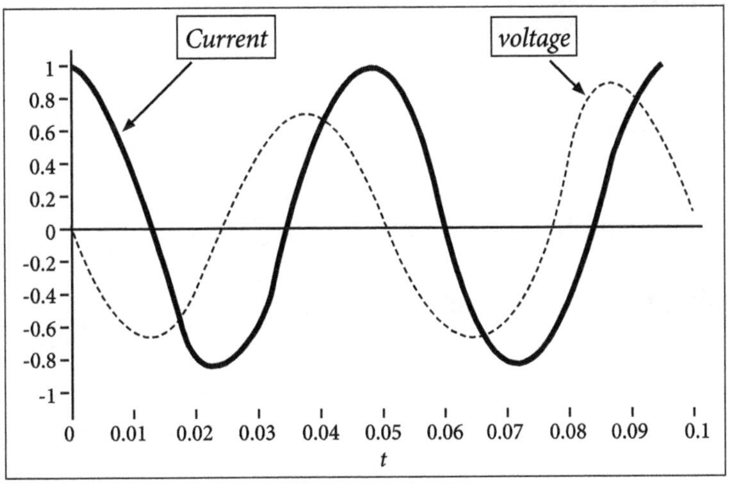

V - I relationship of inductor.

Parallel Combination of Inductors

Inductor connected in series and parallel combines to form an equivalent inductance. Let's first consider the parallel combination of inductors as shown in the below figure. All the inductors have the same voltage across them.

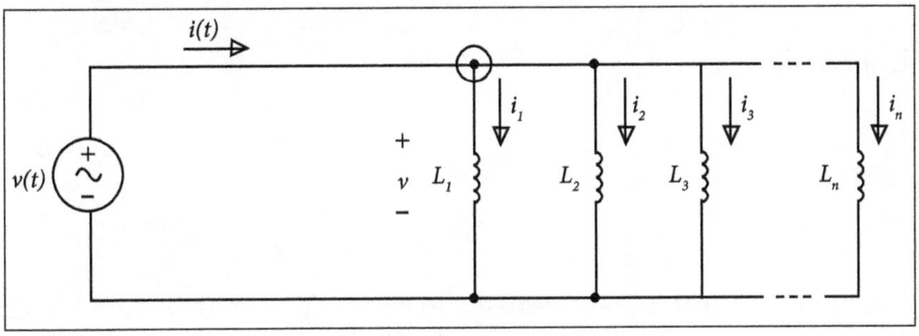

Parallel combination of inductors.

By applying KCL, we can obtain,

$$i = i_1 + i_2 + i_3 + ... + i_n$$

And since $ik = \frac{1}{LK}\int_0^t v\,dt + ik(0)$ we have,

$$i = \frac{1}{L_1}\int_0^t v\,dt + i_1(0) + \frac{1}{L_2}\int_0^t v\,dt + i_2(0) + \frac{1}{L_3}\int_0^t v\,dt + i_3(0) + ... + \frac{1}{L_n}\int_0^t v\,dt + i_n(0)$$

$$=\underbrace{\left(\frac{1}{L_1}+\frac{1}{L_2}+\frac{1}{L_3}+\ldots+\frac{1}{L_n}\right)}_{\frac{1}{L_{eq}}}\int_0^t v\,dt+\underbrace{i_1(o)+i_2(o)+i_3(o)+\ldots+i_n(o)}_{t(o)}$$

$$=\frac{1}{L_{eq}}\int_0^t v\,dt+i(o)$$

Inductors in parallel combines like the resistors in parallel.

Series Combination of Inductors

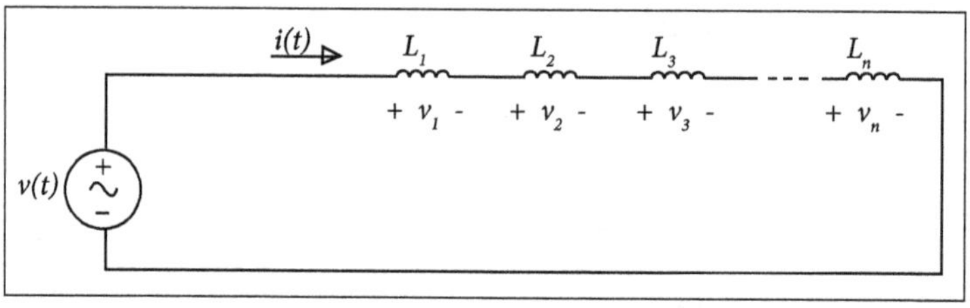

Series combination of inductors.

Now by applying KVL around the loop, we have,

$$v=v_1+v_2+v_3+\ldots.+v_n$$

$$=\underbrace{\left(L_1+L_2+L_3+\ldots+L_n\right)}_{L_{eq}}\frac{di}{dt}$$

$$=L_{eq}\frac{di}{dt}$$

Inductor in series combines like the resistor in series. The energy stored in an inductor is the integral of the instantaneous power delivered to the inductor. Assuming that the inductor had no current flowing through it at $t=-\infty\left[i(-\infty)=o\right]$, then the energy stored in the inductor at time t is given by,

$$E(t)=\int_{-\infty} P(\tau)d\tau$$

$$=\int_{-\infty} v(\tau)i(\tau)d\tau$$

$$=\int_{-\infty} \frac{di(\ )}{}i(\tau)d\tau$$

$$=\frac{1}{2}Li(\tau)^2$$

1.2.3 Capacitance Parameter: Series and Parallel Combination

In both the digital and analog electronic circuits, a capacitor is a fundamental element. It enables the filtering of the signals and it provides a fundamental memory element.

The capacitor is an element which stores energy in the electric field. The circuit symbol and associated electrical variables for the capacitor is shown in the below figure.

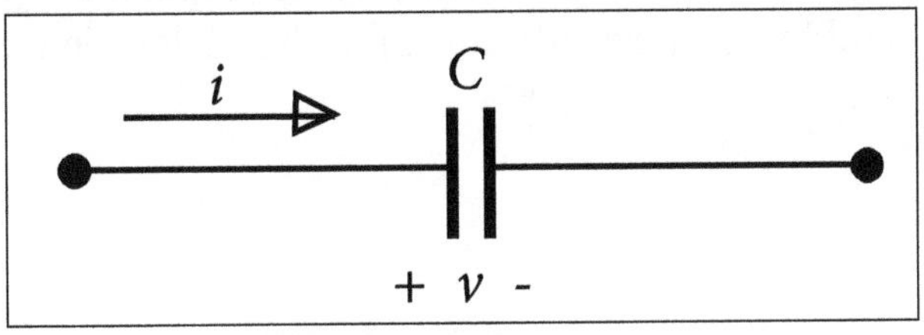

Circuit symbol for capacitor.

The capacitor may be modeled as two conducting plates which are separated by a dielectric as shown in the below given figure. When a voltage v is applied across the plates, a charge +q accumulates on one plate and a charge -q on the other plate.

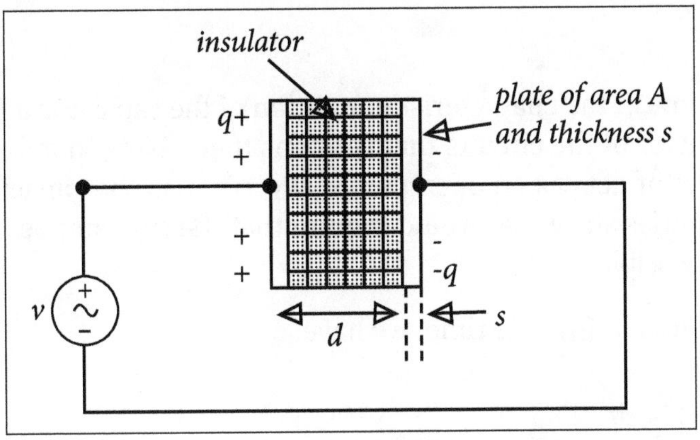

Capacitor model.

If the plates have an area A and are separated by a distance d, the electric field generated across the plates is,

$$E=\frac{q}{\varepsilon A} \quad \text{...(1)}$$

And the voltage across the capacitor plates is given by,

$$v=Ed=\frac{qd}{\varepsilon A} \quad \text{...(2)}$$

The current flowing into the capacitor is the rate of change of charge across the capacitor plates $i=\frac{dq}{dt}$.

Thus, we have,

$$i=\frac{dq}{dt}=\frac{d}{dt}\left(\frac{\varepsilon A}{d}v\right)=\frac{\varepsilon A}{d}\frac{dv}{dt}=C\frac{dv}{dt} \quad ...(3)$$

The constant of proportionality C is referred as the capacitance of the capacitor. It is the function of the geometric characteristics of the capacitor - plate separation (d) and plate area (A) - and by the permittivity (ε) of the dielectric material between the plates,

$$C=\frac{\varepsilon A}{d} \quad ...(4)$$

Capacitance represents the efficiency of the charge storage and it is measured in units of Farads (F).

The current-voltage relationship of the capacitor is given by,

$$i=C\frac{dv}{dt} \quad ...(5)$$

The presence of time in the characteristic equation of the capacitor introduces the new and exciting behavior of the circuits that contains them. Note that for DC signals (dv/dt = 0), the capacitor acts as an open circuit (i = 0). Also the capacitor does not like voltage discontinuities since that would require that the current goes to infinity which is not physically possible.

If we integrate Equation (5) over time, we have,

$$\int_{-\infty}^{t} i\,dt=\int_{-\infty}^{t} C\frac{dv}{dt}dt \quad ...(6)$$

$$v=\frac{1}{C}\int_{-\infty}^{t} i\,dt$$

$$=\frac{1}{C}\int_{0}^{t} i\,dt+v(0) \quad ...(7)$$

The constant of integration v (0) represents the voltage of the capacitor at time t = 0. The presence of the constant of integration v (0) is the reason for the memory properties of the capacitor.

Let us now consider the circuit shown in the below figure where a capacitor of capacitance C is connected with a time varying voltage source v (t).

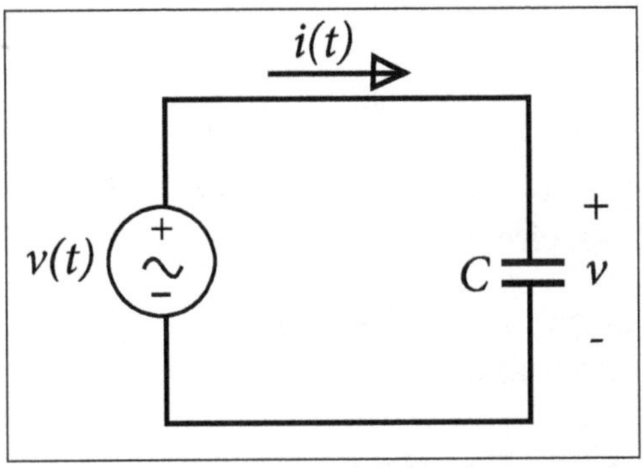

Fundamental capacitor circuit.

If the voltage v (t) has the form

$$v(t)=A\cos(\omega t) \qquad ...(8)$$

Then, the current i(t) becomes,

$$i(t)=C\frac{dv}{dt}$$

$$=-CA\omega\sin(\omega t)$$

$$=C\omega A\cos\left(\omega t+\frac{\pi}{2}\right) \qquad ...(9)$$

Hence the current going through the capacitor and the voltage across the capacitor are 90 degrees out of phase. It is said that the current leads the voltage by 90 degrees. The general plot of the voltage and the current of the capacitor is shown in the figure. The current leads the voltage by 90 degrees.

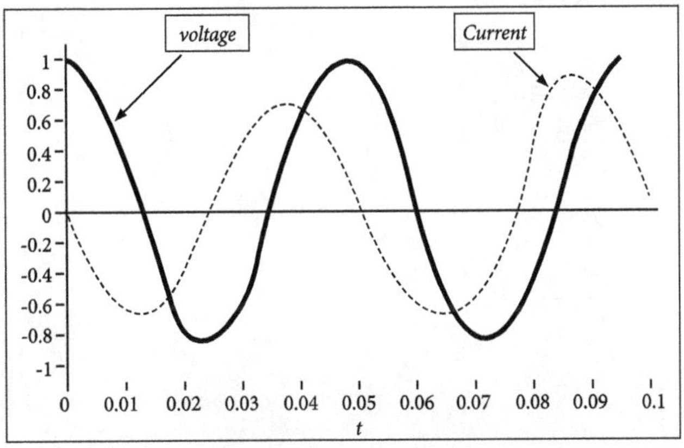

V - I Characteristics of Capacitors.

If we take the ratio of the peak voltage to the peak current we obtain the quantity,

$$X_c = \frac{1}{C\omega} \qquad ...(10)$$

X_c has the units of Volts/Amperes or Ohms and thus it represents some type of resistance. As the frequency $\omega \rightarrow 0$, the quantity X_c goes to infinity which implies that the capacitor resembles an open circuit.

As the frequency becomes very large $\omega \rightarrow \infty$, the quantity X_c goes to zero which implies that the capacitor resembles a short circuit.

Capacitors Connected in Parallel

Capacitors connected in series and in parallel combine to the equivalent capacitance. Let us first consider the parallel combination of the capacitors as shown in the below figure. All the capacitors have the same voltage, v across them.

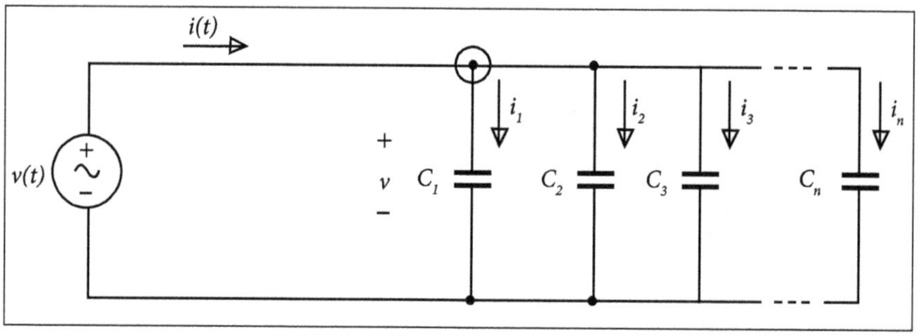

Parallel combination of capacitors.

By applying KCL we obtain,

$$i = i_1 + i_2 + i_3 + ... + i_n \qquad ...(11)$$

and since $ik = Ck\frac{dv}{dt}$ we have,

$$i = C_1\frac{dv}{dt} + C_2\frac{dv}{dt} + C_3\frac{dv}{dt} + ... + C_n\frac{dv}{dt}$$

$$= \left(\underbrace{C_1 + C_2 + C_3 + ... + C_n}_{C_{eq}}\right)\frac{dv}{dt}$$

$$= C_{eq}\frac{dv}{dt} \qquad ...(12)$$

Capacitors Connected in Series

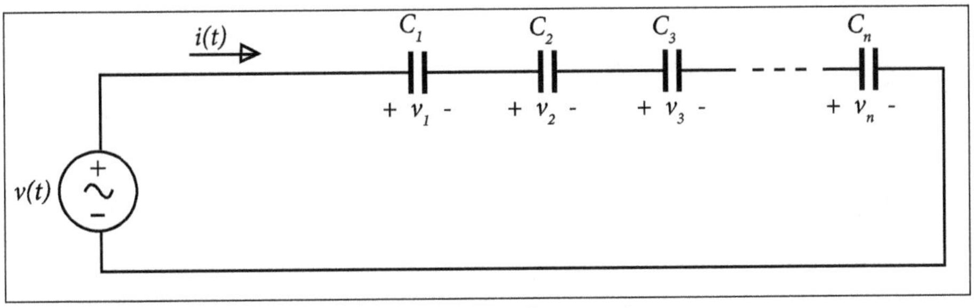

Series combination of n capacitors.

Now by applying KVL around the loop and using Equation (7) we have,

$$v = v_1 + v_2 + v_3 + \ldots + v_n$$

$$= \left(\underbrace{\frac{1}{C_1} + \frac{1}{C_2} + \frac{1}{C_3} + \ldots + \frac{1}{C_n}}_{\frac{1}{C_{eq}}} \right) \int_0^t i(t)dt + v(0)$$

$$= \frac{1}{C_{eq}} \int_0^t i(t)dt + v(0) \qquad \ldots(13)$$

By extension, we can calculate the voltage division rule for the capacitors connected in series. Here, let us consider the case of only two capacitors connected in series as shown in the below figure.

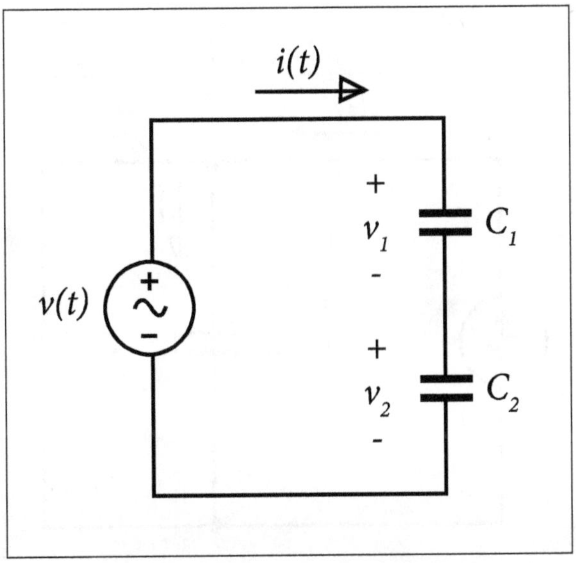

Series combination of two capacitors.

The same current flows through both capacitors and so the voltages v1 and v2 across them are given by,

$$v_1 = \frac{1}{C_1}\int_0^t i\,dt \qquad ....(14)$$

$$v_2 = \frac{1}{C_2}\int_0^t i\,dt \qquad ....(15)$$

And KVL around the loop results in,

$$v(t) = \left(\frac{1}{C_1} + \frac{1}{C_2}\right)\int_0^t i\,dt \qquad ...(16)$$

Which in turn gives the voltages v_1 and v_2 in terms of v and the capacitances.

$$v_1 = v\frac{C_2}{C_1 + C_2} \qquad ...(17)$$

$$v_2 = v\frac{C_1}{C_1 + C_2} \qquad ...(18)$$

Similarly in the parallel arrangement of capacitors the current division rule is given as,

$$i_1 = i\frac{C_1}{C_1 + C_2} \qquad ...(19)$$

$$i_2 = i\frac{C_2}{C_1 + C_2} \qquad ...(20)$$

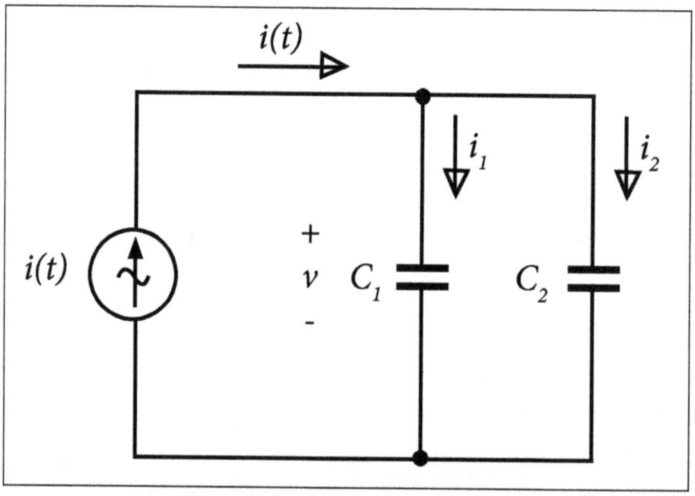

Parallel arrangement of two capacitors.

The instantaneous power delivered to the capacitor is given by,

$$P(t)=i(t)v(t) \quad ...(21)$$

The energy stored in a capacitor is the integral of the instantaneous power. Assuming that the capacitor has no charge across its plates at $t=-\infty\left[v(-\infty)=0\right]$, then the energy stored in the capacitor at time t is given by,

$$E(t)=\int_{-\infty}^{t} P(\tau)d\tau$$

$$=\int_{-\infty}^{t} v(\tau)i(\tau)d\tau$$

$$=\int_{-\infty}^{t} v(\tau)C\frac{dv(\tau)}{d\tau}d\tau$$

$$=\frac{1}{2}C\ v(t)^2 \quad ...(22)$$

If the dielectric material between the plates of the capacitor has the finite resistivity as compared to infinite resistivity in the case of an ideal capacitor, then there is going to be a small amount of current flowing between the capacitor plates. In addition, there are lead resistance and plate effects.

In general, the circuit model of a non-ideal capacitor is shown in the below figure.

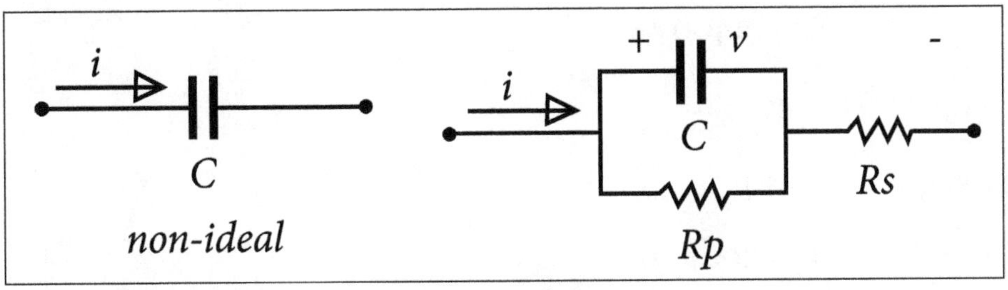

Circuit of non-ideal capacitor.

The resistance R_p is very large and it represents the resistance of the dielectric material. Resistance R_s is typically small and it corresponds to the lead and plate resistance as well as resistance effects due to the operating conditions.

In practice, we are concerned with in series resistance of a capacitor called the Equivalent Series Resistance (ESR). ESR is a very important capacitor characteristic and must be taken into consideration in circuit design. Therefore, the non-ideal capacitor model is shown in the below figure.

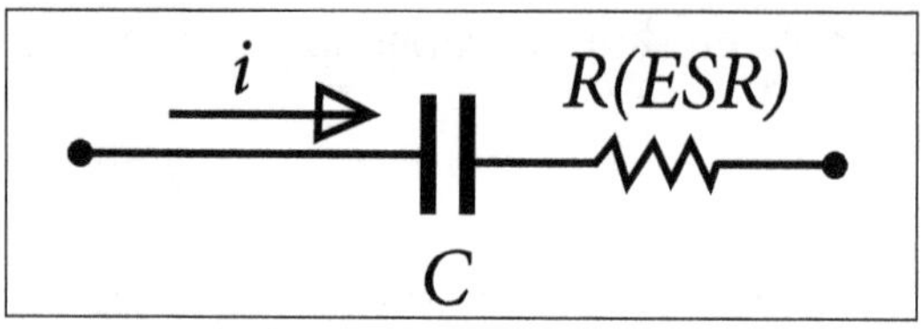

Non-ideal capacitor with series resistor.

Typical values of ESR is in the mΩ-Ω range.

Problems

1. The effective resistance of two resistors connected in series is 100 Ω. When connected in parallel, its effective value is 24 ohms. Let us determine the value of two resistors.

Solution:

Given:

Series $R_1 + R_2 = 100\ \Omega$ => $R_2 = 100 - R_1$

$$R_1R_2 / R_1 + R_2 = 24$$

$$R_1R_2 / 100 = 24$$

$$R_1R_2 = 2400$$

$$R_1(100 - R_1) = 2400$$

$$100\ R_1 - R_1^2 - 2400 = 0$$

$$R_1^2 - 100\ R_1 + 2400 = 0$$

$$(R_1 - 60)(R_1 - 40) = 0$$

Therefore, $R_1 = 60$ (or) $R_1 = 40$

When $R_1 = 60\ \Omega$; $R_2 = 100 - 60 = 40\ \Omega$

When $R_1 = 40\ \Omega$; $R_2 = 100 - 40 = 60\ \Omega$.

2. Let us determine the R_{eq} between two points A & B.

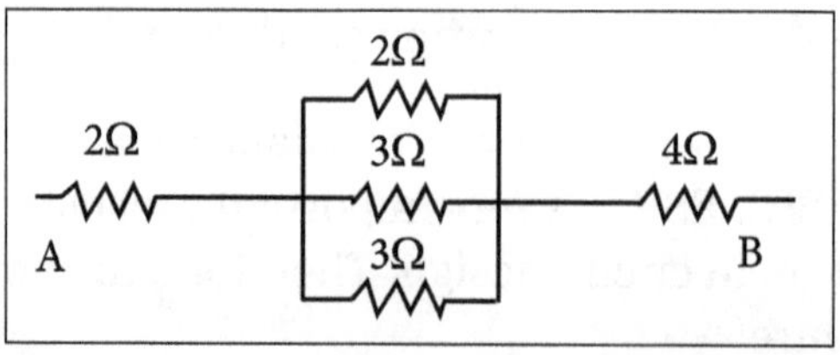

Solution:

Given:

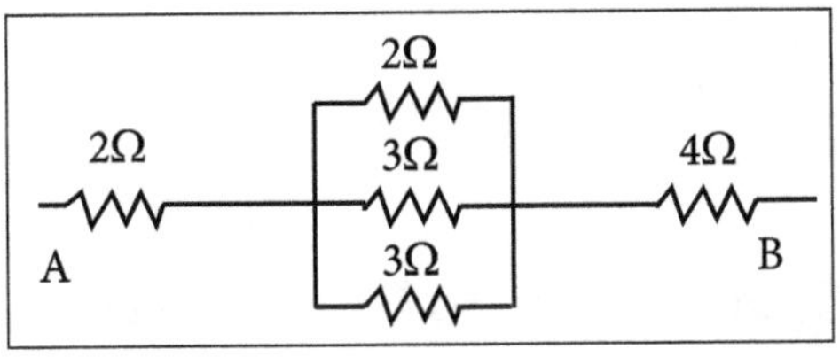

$$1/R_{eq} = ½+1/3+1/3=1.17\ (R_{eq} = 1/1.17 = 0.8547)$$

$$R_{eq} = 0.854$$

3. A 47μF capacitor is connected to a voltage which varies in time as v(t) = 20sin (200πt) volts. Let us calculate the current i(t) through the capacitor.

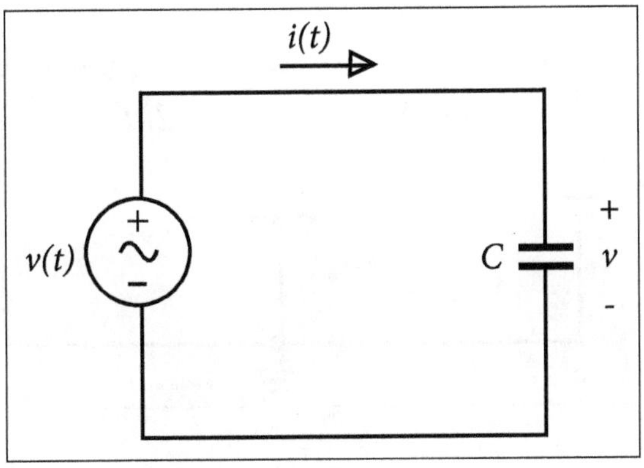

Solution:

Given:

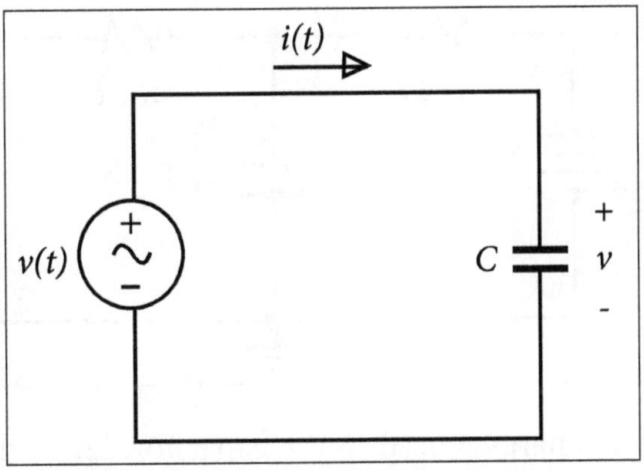

$$v(t)=20\sin(200\pi t)\ \text{volts}$$

$C = 47\,\mu F$

The current is given by,

$$i = C\frac{dv}{dt}$$

$$= 47\times10^{-6}\frac{d}{dt}20\sin(200\pi t) = 47\times10^{-6}\times20\times200\pi\cos(200\pi t)$$

$$= 0.59\cos(200\pi t)\,\text{Amperes}$$

4. Let us calculate the energy stored in the capacitor of the circuit to the right under DC conditions.

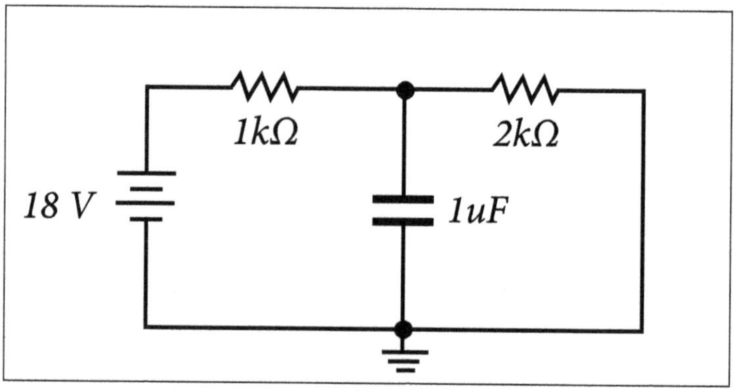

Solution:

Given:

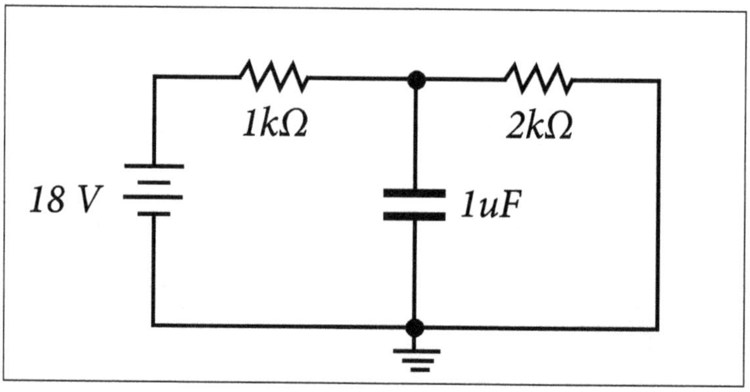

In order to calculate the energy stored in the capacitor, let us determine the voltage across it. Under DC conditions, the capacitor appears as an open circuit (no current flowing through it). Thus, the corresponding circuit is as below.

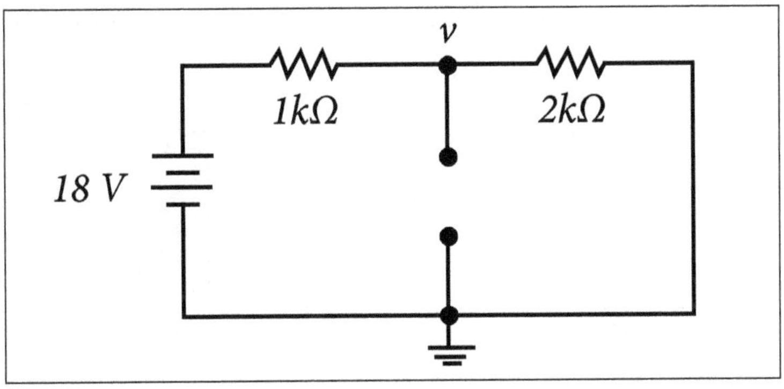

And from the voltage divider formed by the 1kΩ and the 2kΩ resistors, the voltage v is 12Volts. Thus, the energy stored in the capacitor is given by,

$$Ec = \frac{1}{2}Cv^2 = \frac{1}{2}1\times10^{-6}\times12^2 = 72\,\mu\,\text{Joules}$$

1.3 Energy Sources: Ideal, Non-ideal, Independent and Dependent Sources

Ideal Voltage Sources

An active two-terminal element that supplies energy to the circuit is a source of energy. An ideal voltage source is a circuit element which maintains the prescribed voltage across the terminals regardless of the current flowing in those terminals.

Similarly, an ideal current source is the circuit element that maintains the prescribed current through its terminals, regardless of the voltage across those terminals. These circuit elements do not exist as practical devices; they are the only idealized models of actual voltage and current sources.

The ideal voltage and current sources can be further classified as either independent sources or dependent sources. The independent source establishes the voltage or current in a circuit without relying on the voltages or currents elsewhere in the circuit. The value of the voltage or current supplied is specified by the value of the independent source alone.

The dependent source establishes the voltage or current whose value depends on the value of the voltage or current elsewhere in the circuit. We cannot specify the value of the dependent source, unless we know the value of the voltage or current on which it depends.

The circuit symbols for ideal independent sources are shown in figure (a) and (b). Circle is used to represent the independent source. The circuit symbols for dependent sources are shown in figure(c), (d), (e) and (f). Diamond symbol is used to represent a dependent source.

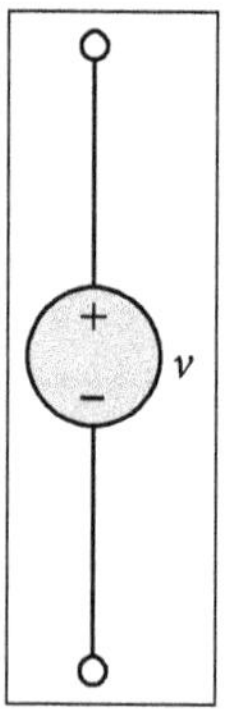

(a) An ideal independent voltage source.

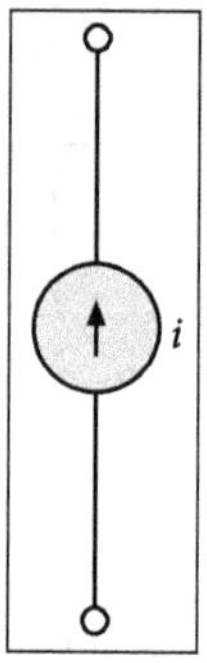

(b) An ideal independent current source.

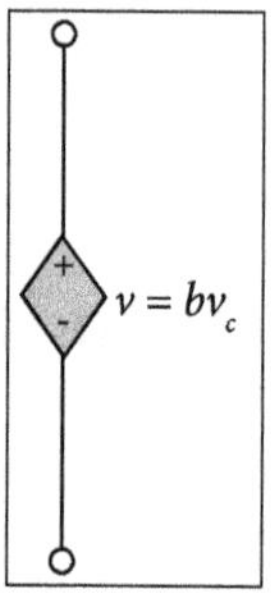

(c) Voltage controlled voltage source.

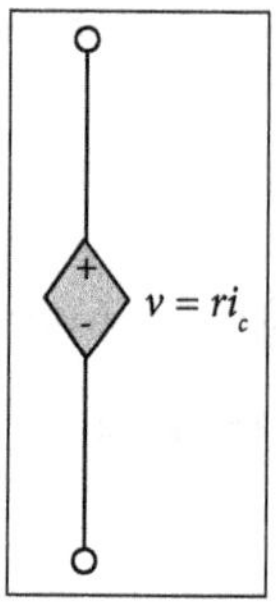

(d) Current controlled voltage source.

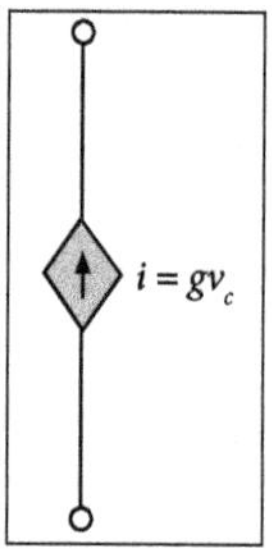

(e) Voltage controlled current source.

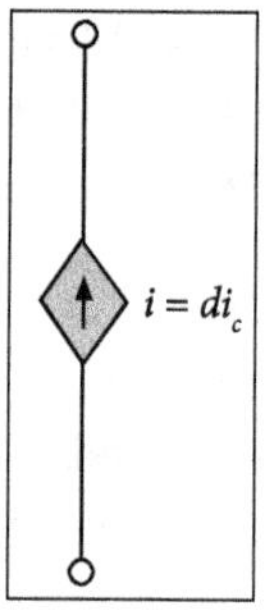

(f) Current controlled current source.

1.3.1 Non-ideal, Independent and Dependent Sources

Non-Ideal Voltage Sources

The non-ideal voltage source delivers a voltage V and a current i,

Where,

$$V = V_s - i \cdot R_s \qquad ...(1)$$

The above equation(1) indicates that the voltage delivered by our non-ideal voltage source model decreases as the current out of the voltage source increases, which agrees with expectations.

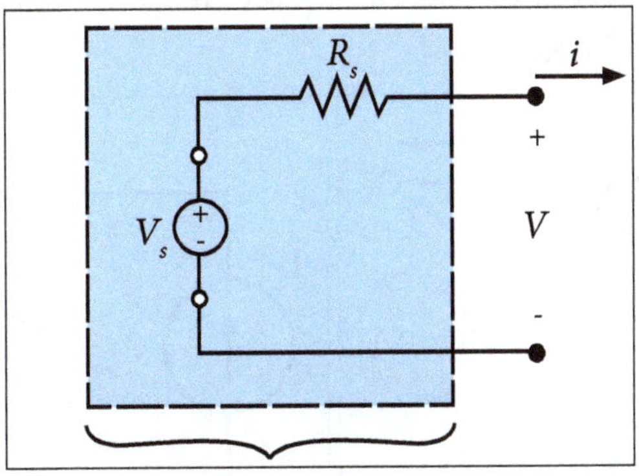

Non-ideal voltage source model.

Non-Ideal Current Sources

An ideal current source was providing a specified current, regardless of the voltage difference across the device. This model suffers from the same basic drawback as our ideal voltage source model. The model can deliver infinite power, which is inconsistent with the capabilities of a real current source. The source delivers a voltage V and current i. The output current is given by,

$$i = i_s - \frac{V}{R_S} \qquad ...(2)$$

The above equation (2) shows that the current delivered by the source decreases as the delivered voltage increases.

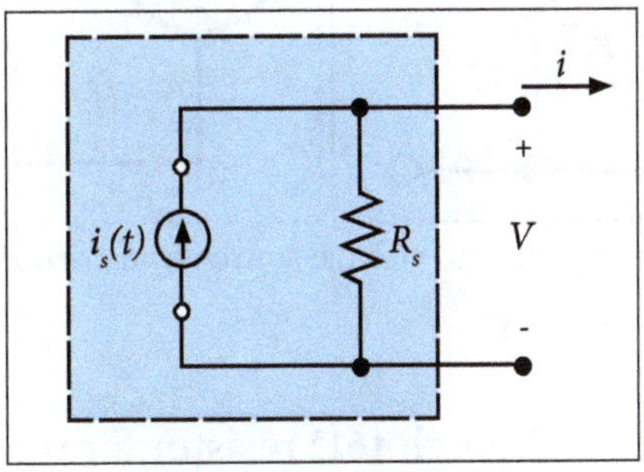

Non-ideal current source model.

1.4 Source Transformation

Source transformation is defined as to concert the sources for easy analysis of circuit. In mesh analysis, it is easier if the circuit has voltage sources. In nodal analysis, it is easier if the circuit has current sources.

Voltage Source to Current Source Transformation

If voltage source is converted to current source, then the current source $I=\frac{V}{R_{se}}$ with parallel resistance equal to R_{se}.

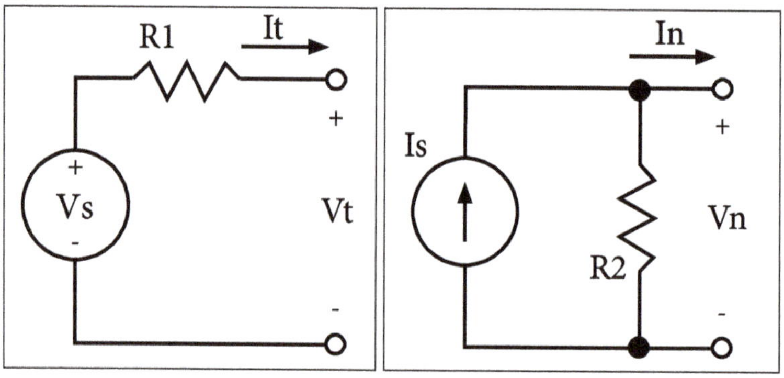

Voltage source to current source transformation.

Current Source to Voltage Source Transformation

If current source is converted to voltage source, then the voltage source $V = IR_{sh}$ with series resistance equal to R_{sh}.

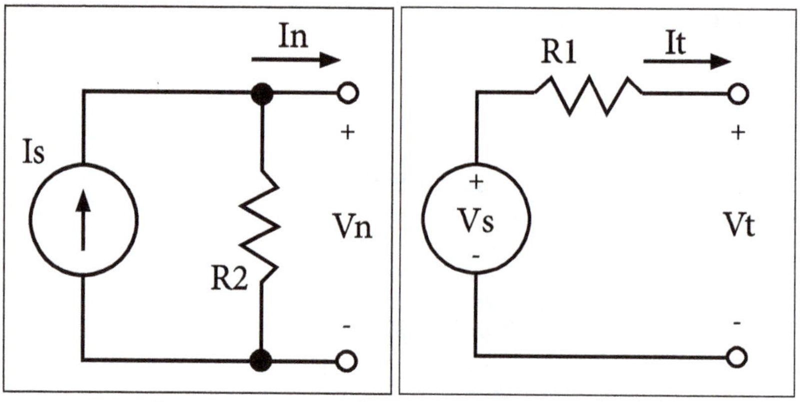

Current source to voltage source transformation.

Problems

1. Let us determine the current through 16Ω resistor for the given diagram.

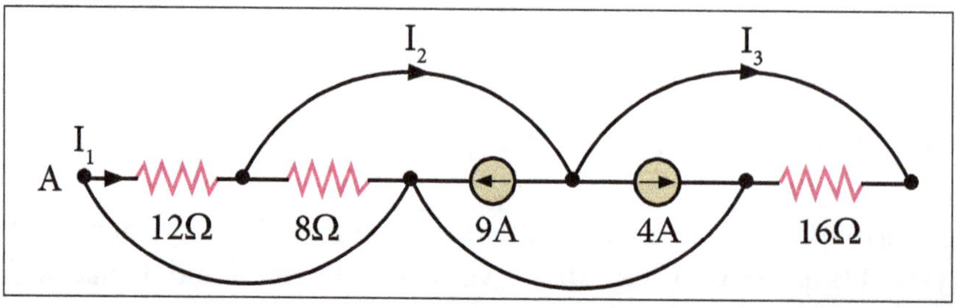

Solution:

Assigning currents to each branch using KCL,

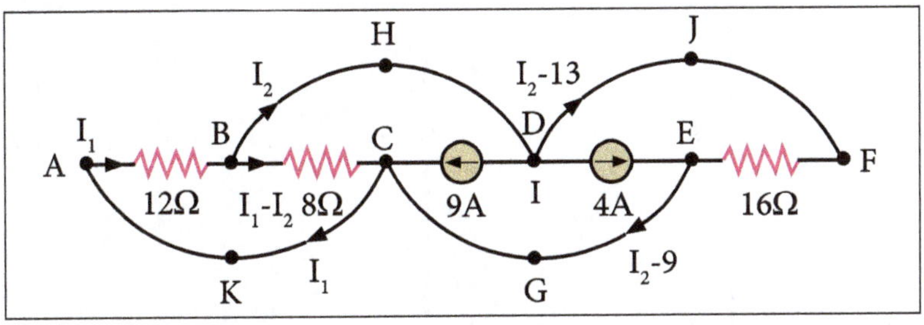

By applying KVL to ABCKA,

$$-12\,I_1 - 8(I_1 - I_2) = 0$$

$$-20\,I_1 + 8\,I_2 = 0 \qquad \ldots(1)$$

By applying KVL to ABHDJFEGCKA,

$$-12\,I_1 - 16\,(I_2 - 1_3) = 0$$

$$-12\,I_1 - 16\,I_2 = -\,208 \qquad \ldots(2)$$

Solving Equations (1) and (2), we get,

$$I_1 = 4\text{ A }(\rightarrow)$$

$$I_2 = 10\text{ A }(\curvearrowright)$$

$$I_{16\,\Omega} = I_2 - 1_3$$

$$= 10 - 13 = -\,3\text{A}$$

$$I_{16\,\Omega} = 3\text{A}$$

Result:

Current through 16Ω resistor, $I_{16\,\Omega} = 3\text{A}$

2. Let us determine the equivalent voltage source for the current source shown in the below figure.

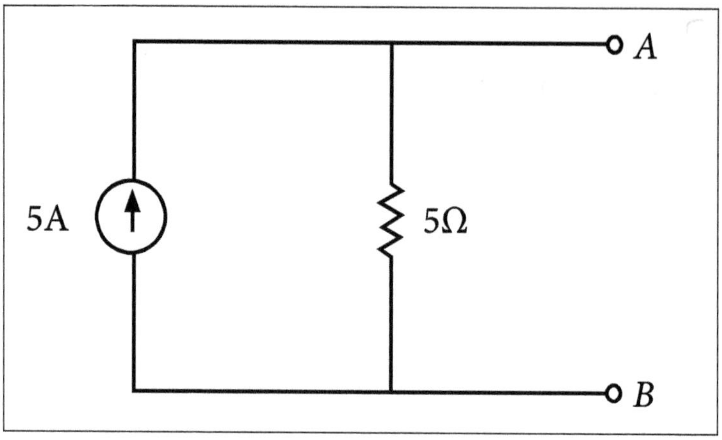

Solution:

The voltage across terminals A and B is equal to 25 V. Since the internal resistance for the current source is 5 Ω, the internal resistance of the voltage source is also 5 Ω. The equivalent voltage is shown in the below figure.

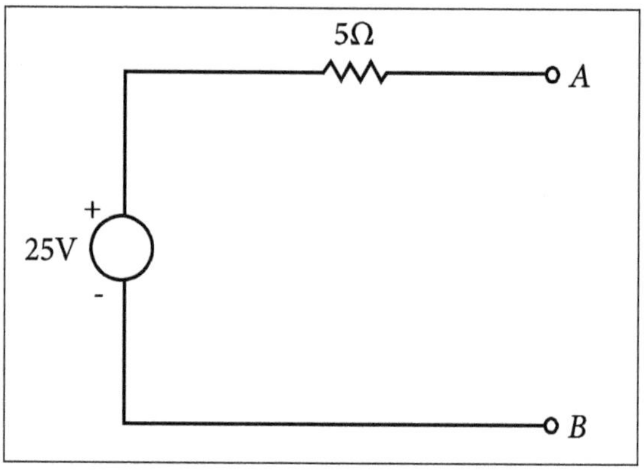

3. Let us determine the equivalent voltage source for the current source shown in the below figure.

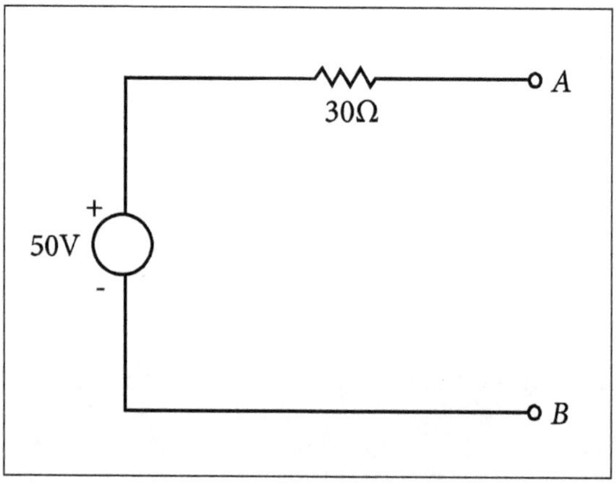

Solution:

The short-circuit current at terminals A and B is equal to,

$$I=\frac{50}{30}=1.66\,A$$

Since the internal resistance for the voltage source is 30 Ω, the internal resistance of the current source is also 30 Ω. The equivalent current source is shown in the below figure.

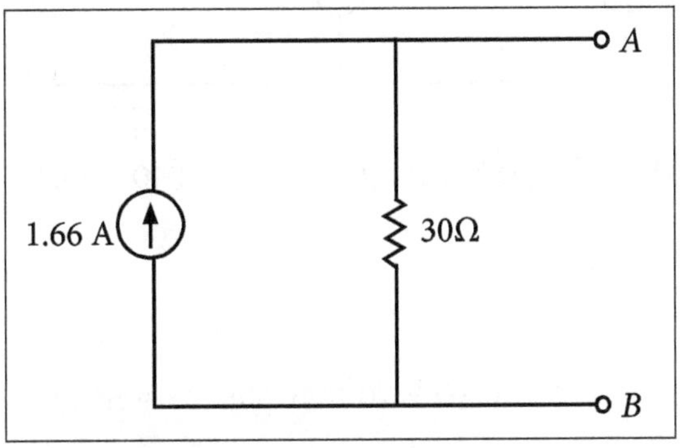

4. Using source transformation, let us find the power delivered by the 50 V voltage circuit as shown in the below figure.

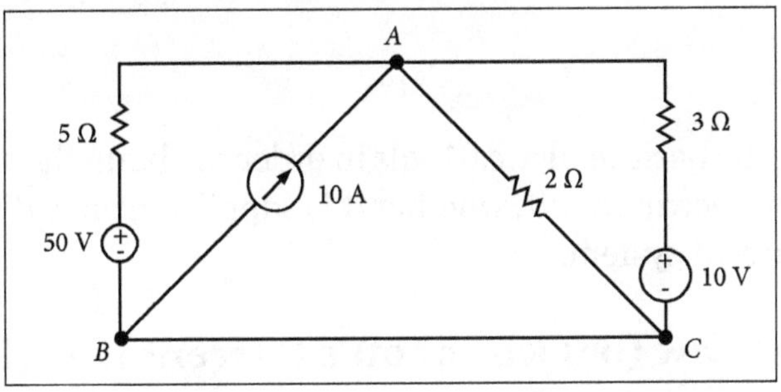

Solution:

The current source in the circuit in the above figure can be replaced by a voltage source as shown in the below figure.

$$\frac{V-50}{5}+\frac{V-20}{2}+\frac{V-10}{3}=0$$

$$V[0.2+0.5+0.33]=23.33$$

or,

$$V=\frac{23.33}{1.03}=22.65\text{ V}$$

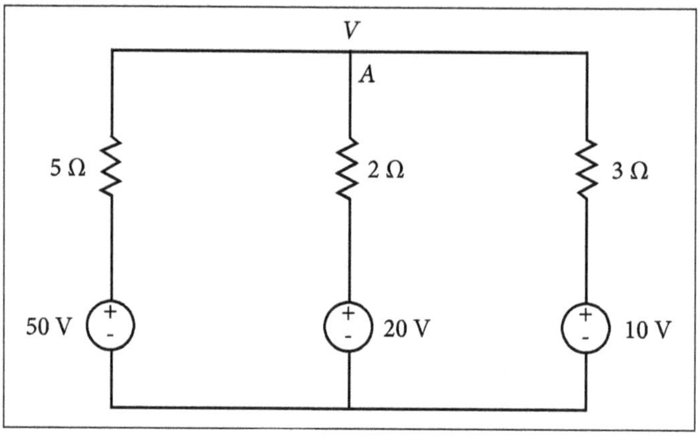

∴ The current delivered by the 50 V voltage source is (50 - V)/5,

$$= \frac{50 - 22.65}{5} = 5.47\,\text{A}$$

Hence, the power delivered by the 50 V voltage source $= 50 \times 5.47 = 273.5$ W.

1.5 Kirchhoff's Laws, Mesh Analysis and Nodal Analysis

Kirchhoff's Law

Kirchhoff's laws are the basic analytical tools in order to obtain the solutions of voltages and currents for any electric circuit, whether it is supplied from a direct current system or an alternating current system.

Kirchhoff's First Law (or) Kirchhoff's Current Law (KCL)

"The total charge or current entering a junction or node is exactly equal to the charge leaving the node as it has no other place to go except to leave, as no charge is lost within the node".

In other words, KCL states that at any node (junction) in a circuit, the algebraic sum of currents entering and leaving a node at any instant of time must be equal to zero,

$$I_{(exiting)} + I_{(entering)} = 0$$

This idea by Kirchhoff is known as the Conservation of Charge.

At any instant of time, the algebraic sum of current at a junction or node is zero. Otherwise, the algebraic sum of current entering into a junction is equal to the current leaving the junction.

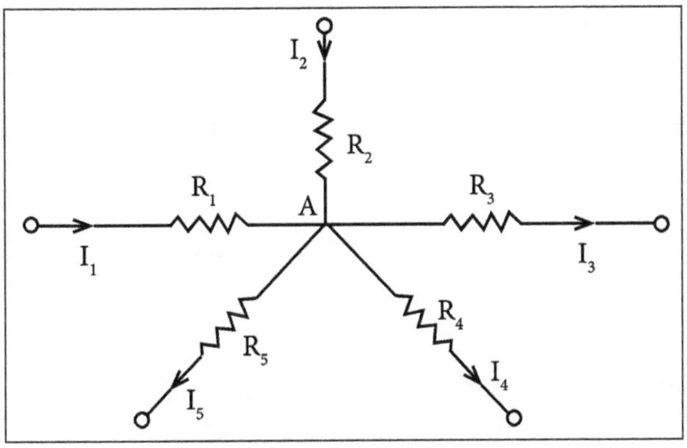

Kirchhoff's current law.

$$\Rightarrow I_1 + I_2 = I_3 + I_4 + I_5 \text{ (or) } I_1 + I_2 - I_3 - I_4 - I_5 = 0.$$

Kirchhoff's Second Law (or) Kirchhoff's Voltage Law (KVL)

"In any closed loop network, the total voltage around the loop is equal to the sum of all the voltage drops within the same loop which is also equal to zero".

In other words, KVL states that in a closed circuit, the algebraic sum of all the source voltages must be equal to the algebraic sum of all the voltage drops. This idea by Kirchhoff is known as the Conservation of Energy.

The algebraic sum of potential around a closed path is equal to zero. In other words, the sum of potential rise is equal to the sum of potential drop in a closed circuit.

$$V = V_1 + V_2 + V_3 \text{ (or) } V - V_1 - V_2 - V_3 = 0$$

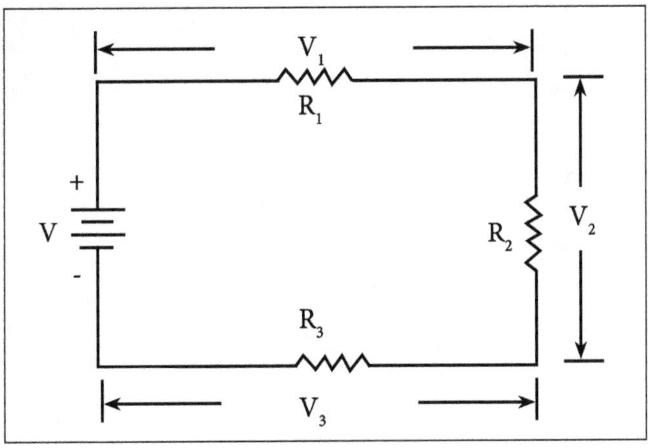

Kirchhoff's voltage law.

Steps to solve a problem using KCL / KVL method:

- Step 1: Find the branch currents of each branch using KCL.

- Step 2: Using KVL, get the equations for each loop in terms of unknown currents.
- Step 3: Solve the simultaneous equations obtained in step 2 using calculator to get the unknown currents.
- Step 4: Find the required branch current and power if required.

Problems

1. Let us consider a current of 0.55A is flowing through the resistance of 10 Ω. Let us determine the potential difference between its ends.

Solution:

Given:

Current, $I = 0.55A$

Resistance, $R = 10\Omega$

Formula to be used:

$V = IR$

$= 0.55 \times 10$

$= 5.5\ V$

2. Let us consider a supply voltage of 200 V is applied to a 100 Ω resistor. Let us determine the current flowing through it.

Solution:

Given:

Voltage, $V = 200V$

Resistance, $R = 100\Omega$

Formula to be used:

Current, $I = V / R$

$= 200/100$

$= 2\ A$

3. Let us calculate the resistance of the conductor if a current of 2 A flows through it when the potential difference across its ends is 12 V.

Solution:

Given:

Current, $I = 2A$

Potential difference, $V = 12$

Formula to be used:

Resistance, $R = V/I$

$= 12/2$

$= 6$ ohm

4. Let us determine the current flowing in the 40 Ω Resistor, R_3.

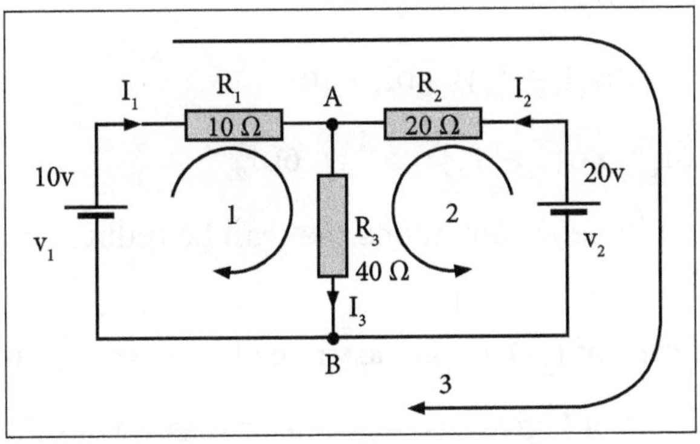

Solution:

Given:

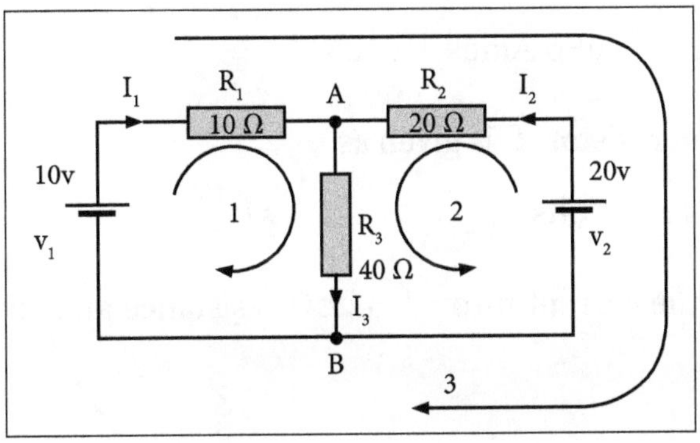

The circuit has three branches, two nodes (A and B) and two independent loops.

Formula to be used:

Using Kirchhoff's Current Law, KCL, the equations are given as,

$$\text{At node A: } I_1 + I_2 = I_3$$

$$\text{At node B: } I_3 = I_1 + I_2$$

Using Kirchhoff's Voltage Law, KVL, the equations are given as,

$$\text{Loop 1 is given as, } 10 = R_1 \times I_1 + R_3 \times I_3 = 10I_1 + 40I_3 \quad ...(1)$$

$$\text{Loop 2 is given as, } 20 = R_2 \times I_2 + R_3 \times I_3 = 20I_2 + 40I_3 \quad ...(2)$$

$$\text{Loop 3 is given as, } 10 - 20 = 10I_1 - 20I_2 \quad ...(3)$$

As,

I_3 is the sum of $I_1 + I_2$

We can rewrite the equations as,

Equation (1): $10 = 10I_1 + 40(I_1 + I_2) = 50I_1 + 40I_2$

Equation (2): $20 = 20I_1 + 40(I_1 + I_2) = 40I_1 + 60I_2$

We now have two simultaneous equations that can be reduced to give us the value of both I_1 and I_2.

Substitution of I_1 in terms of I_2 gives us the value of I_1 as -0.143 Amps.

Substitution of I_2 in terms of I_1 gives us the value of I_2 as +0.429 Amps.

$$\text{As, } I_3 = I_1 + I_2$$

The current flowing in resistor R_3 is given as,

$$-0.143 + 0.429 = 0.286 \text{ Amps}$$

The voltage across the resistor R_3 is given as,

$$0.286 \times 40 = 11.44 \text{ volts}$$

5. Let us determine the current through a 20Ω resistance and current through a 40Ω resistance.

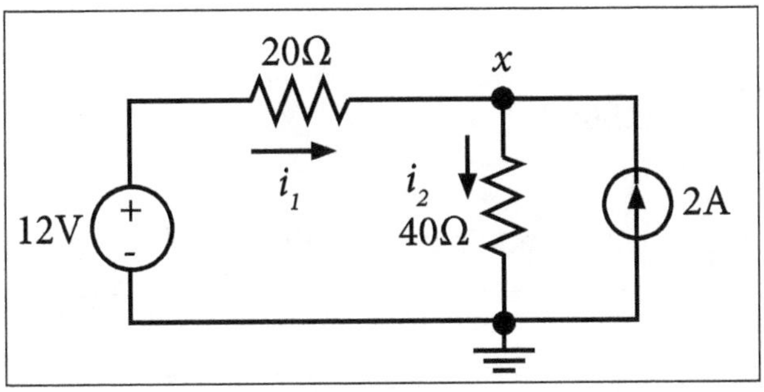

Solution:

Given:

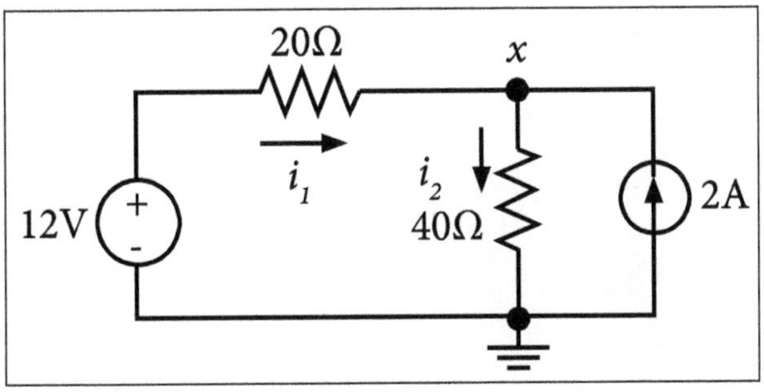

By writing KCL at node x,

$$-i_1 + i_2 - 2A = 0$$

Write v_x in the circuit using Ohm's Law,

$$i_1 = \frac{12V - v_x}{20\Omega}, \quad i_2 = \frac{v_x}{40\Omega}$$

Apply last two equation into KCL at node x,

$$-i_1 + i_2 - 2A = -\frac{12V - v_x}{20\Omega} + \frac{v_x}{40\Omega} - 2A = 0$$

$$-0.6 + 0.05\, v_x + 0.025\, v_x - 2 = 0$$

$$v_x = 34.67V$$

The current through a 20Ω resistance,

$$i_1 = \frac{12V - 34.67}{20\Omega} = -1.133\,A$$

The current through a 40Ω resistance,

$$i_2 = \frac{v_x}{40\Omega} = \frac{34.67}{40\Omega} = 0.866\,A$$

6. Let us find the current in a circuit using Kirchhoff's voltage law.

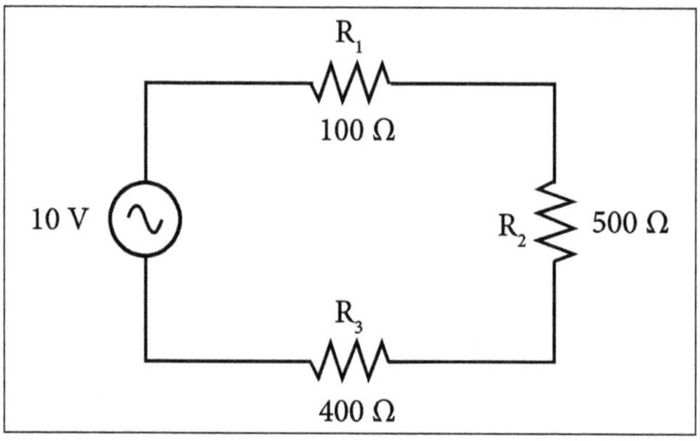

Solution:

Given:

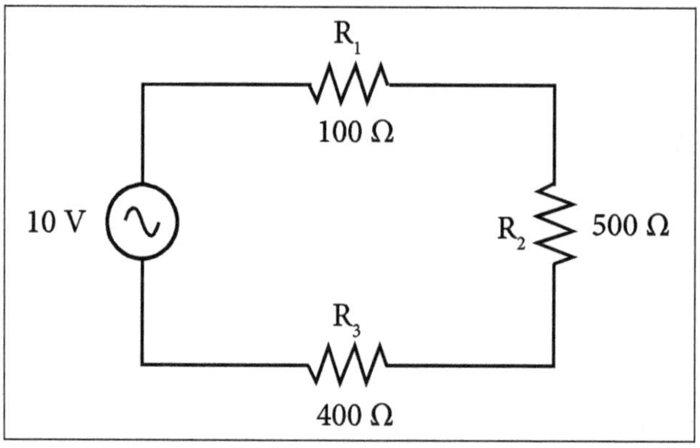

Formula:

$$V = IR$$

By applying KVL equations for voltages,

$$10 = 100(I) + 500(I) + 400(I)$$

$$10 = 1000(I)$$

$$10 = 1000(I) = I = 10/1000$$

$$I = 0.01A$$

7. Let us determine the current i and voltage v over each resistor.

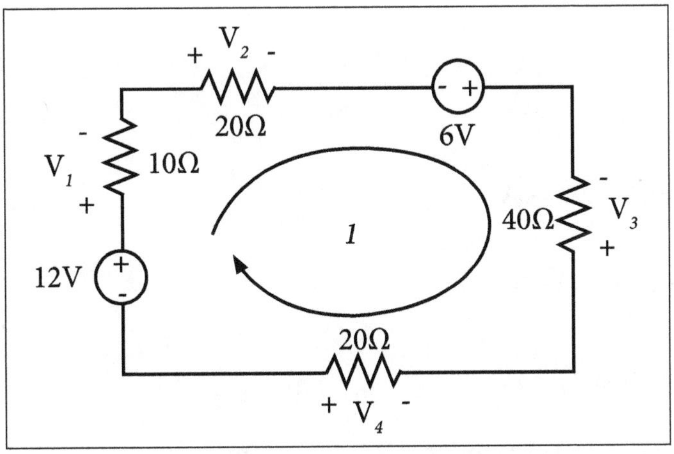

Solution:

Given:

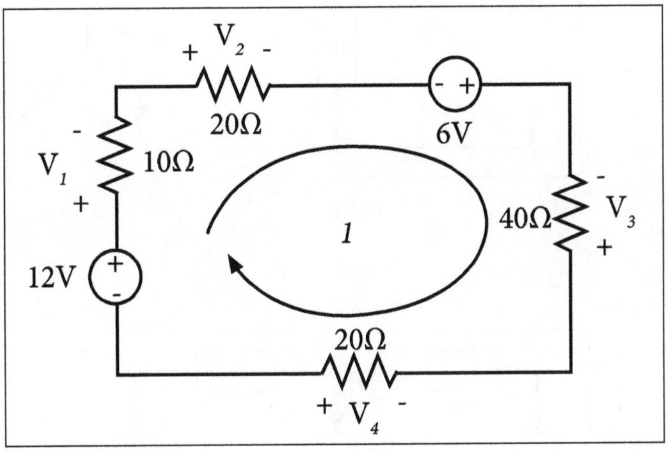

By applying KVL equations for voltages,

$$v_1 + v_2 + v_3 + v_4 = 18$$

Using Ohm's Law,

$$v_1 = 10\Omega,\ v_2 = 20\Omega,\ v_3 = 40\Omega,\ v_4 = 20\Omega$$

Substituting into KVL equation, we get,

$$10i + 20i + 40i + 20i = 18$$

$$90i = 18$$

$$i = \frac{18}{90} = 0.2\,A$$

$$v_1 = R_1 i = 10(0.2) = 2V$$

$$v_2 = R_2 i = 20(0.2) = 4V$$

$$v_3 = R_3 i = 40(0.2) = 8V$$

$$v_4 = R_4 i = 20(0.2) = 4V$$

8. Let us determine v_1 and v_2 in the following circuit.

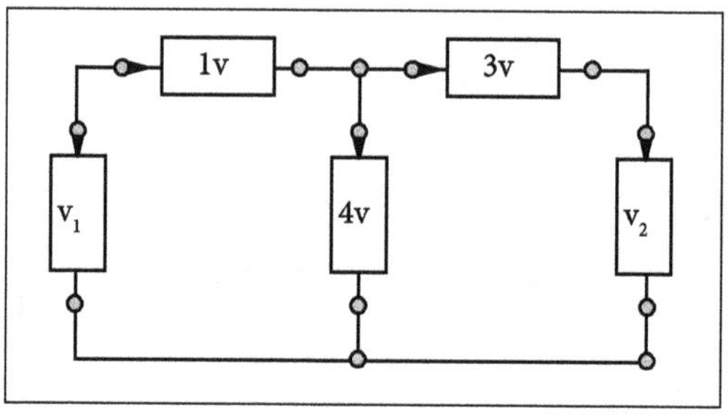

Solution:

Given:

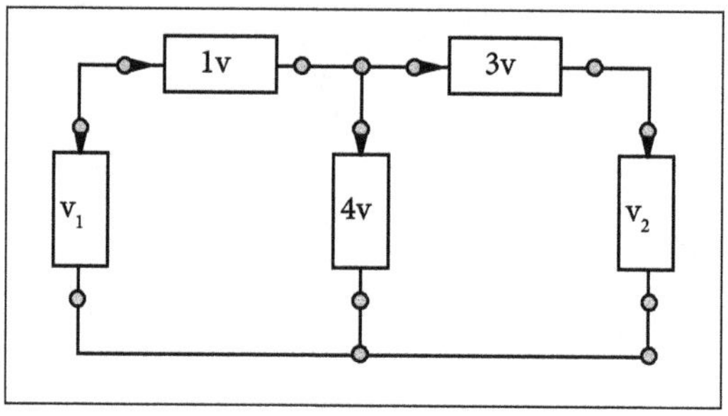

Loop 1:

$$-V_1 + 1 + 4 = 0$$

$$V_1 = 5V$$

Loop 2:

$$-4 + 3 + V_2 = 0$$

$V_2 = 1V$

9. Let us determine V_1, V_2 and V_3 where the arrows are signifying the positive position of the box and the negative is at the end of the box.

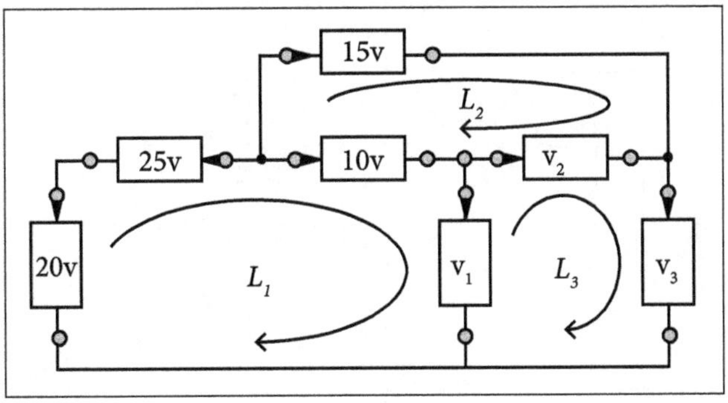

Solution:

Given:

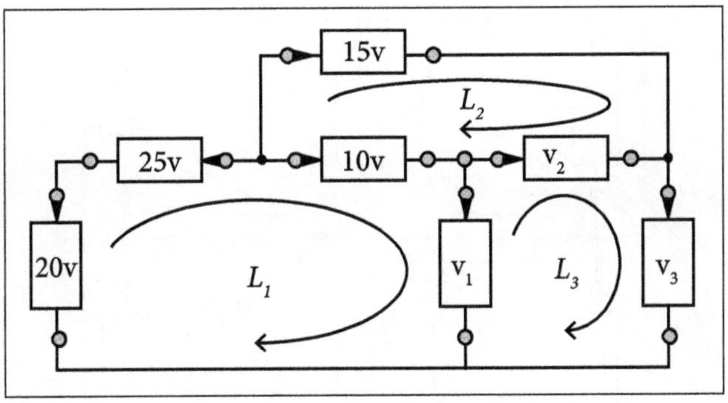

Loop 1:

$$-20 - 25 + 10 + V_1 = 0$$

$$V_1 = 35V$$

Loop 2:

$$-10 + 15 - V_2 = 0$$

$$V_2 = 5V$$

Loop 3:

$$-V_1 + V_2 + V_3 = 0$$

$$-35 + 5 + V_3 = 0$$

$$V_3 = 30V$$

10. Let us determine V_1, V_2, V_3 and V_4 from the below figure in which the arrows are signifying the positive position of the box and the negative is at the end of the box.

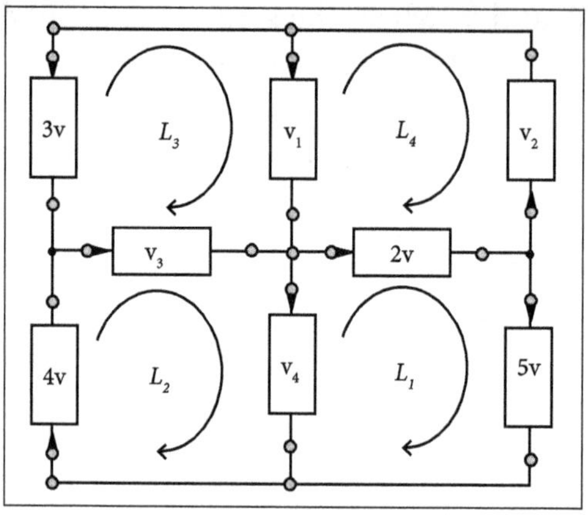

Solution:

Given:

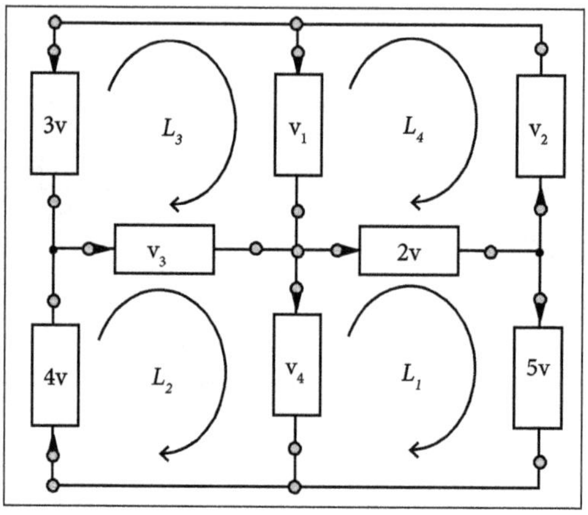

Loop 1:

$$-V_4 + 2 + 5 = 0$$

$$V_4 = 7V$$

Loop 2:

$$4 + V_3 + V_4 = 0$$

$$V_3 = -4 - 7$$

$$V_3 = -11V$$

Loop 3:

$$-3+V_1-V_3=0$$

$$V_1=V_3+3$$

$$V_1=-8V$$

Loop 4:

$$-V_1-V_2-2=0$$

$$V_2=-V_1-2$$

$$V_2=6V$$

11. Let us determine V_1, V_2 and V_3 in the following circuit.

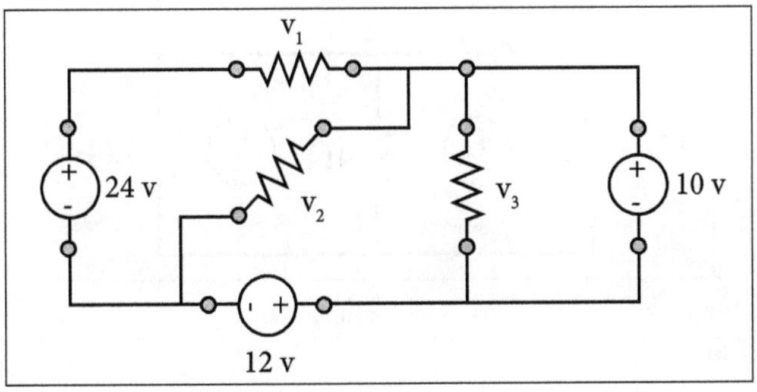

Solution:

Given:

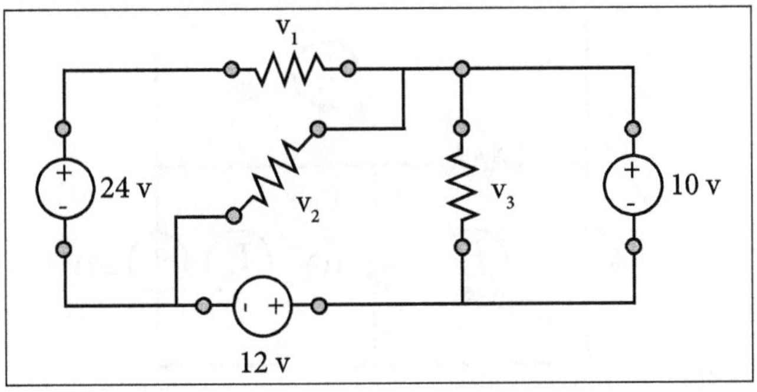

Outer loop:

$$-24+V_1+10+12=0$$

$$V_1=2V$$

V_2 and 10V source loop:

$$V_2 + 10 + 12 = 0$$

$$V_2 = -22V$$

V_3 and 10V source loop:

$$-V_3 + 10 = 0$$

$$V_3 = 10V$$

12. Let us determine I_1, I_2, I_3 in the following circuit.

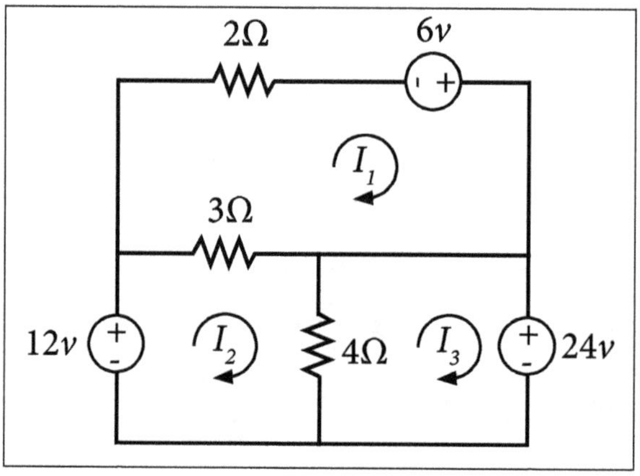

Solution:

Given:

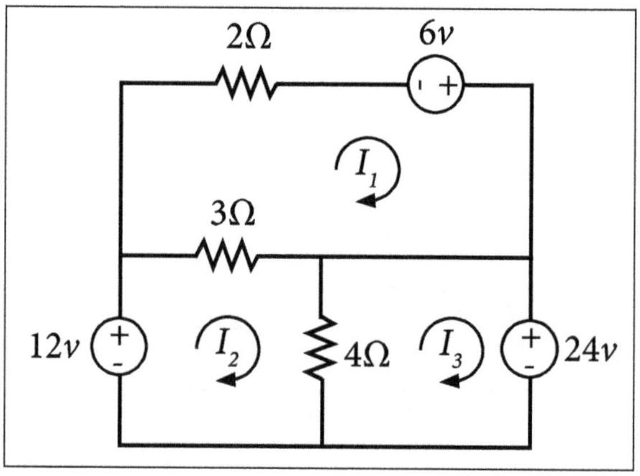

Loop 1:

$$-6 = -2\,I_1 + 3(I_2 - I_1)$$

Loop 2:

$$12 = 3(I_2 - I_1) + 4(I_2 - I_3)$$

Loop 3:

$$24 = -4(I_3 - I_2)$$

From Loop 3:

$$(I_2 - I_3) = 6$$

Substitute in loop 2, the results is,

$$12 = 3(I_2 - I_1) + 4(6)$$

$$I_2 - I_1 = -4$$

Substitute in loop 1, the results is,

$$-6 = -2I_1 + 3(-4)$$

$$I_1 = -3A$$

Substitute in loop 2, the results is,

$$I_2 - (-3A) = -4A$$

$$I_2 = -7A$$

Substitute in loop 3, the results is,

$$24 = -4\ (I_3 + 7)$$

$$-6 = I_3 + 7$$

$$I_3 = -13A$$

1.5.1 Mesh Analysis and Nodal Analysis

Mesh Current Method

A 'mesh' is any closed path in a given circuit that does not have any element (or branch) inside it. A mesh has the properties such as:

- Every node in the closed path is exactly formed with two branches.

- No other branches are enclosed by the closed path.

Mesh Analysis

This is an alternative structured approach for solving the circuit and is based on calculating mesh currents. A similar approach to the node situation is used. A set of equations (based on KVL for each mesh) is formed and the equations are solved for unknown values. As many equations are needed as unknown, mesh currents exist.

Steps for analyzing the mesh current:

- Step 1: Identify the mesh currents.
- Step 2: Determine which mesh currents are known.
- Step 3: Write equation for each mesh using KVL and that includes the mesh currents.
- Step 4: Then, solve the equations.

Node Voltage Method

In the node voltage method, we identify all the nodes on the circuit. Choosing one of them as the reference voltage (i.e., zero potential) and subsequently assign other node voltages (unknown) with respect to the reference voltage.

If the circuit has 'n' nodes, there are 'n-1' node voltages are unknown. At each of these 'n-1' nodes, we can apply KCL equation. The unknown node voltages become the independent variables of the problem and the solution of node voltages can be obtained by solving a set of simultaneous equations.

Nodal Analysis

Nodal analysis involves looking at a circuit and determining all the node voltages in the circuit. The voltage at any given node of circuit is the voltage drop between that node and a reference node (usually ground). Firstly, all the nodes in the circuit are counted and identified. Secondly, nodes at which the voltage is already known are listed.

A set of equations based on the node voltages are formed and these equations are solved for unknown quantities. The set of equations are formed using the KCL equations at each node. The set of simultaneous equations that is produced is then solved. Then, branch currents can be found once the node voltages are known. This can be reduced to a series of steps:

- Step 1: Identify the nodes.
- Step 2: Choose a reference node.

- Step 3: Identify which node voltages are known if any.
- Step 4: Identify the branch currents.
- Step 5: Use KCL to write an equation for each unknown node voltage.
- Step 6: Then, solve the equations.

Problems

1. For the circuit shown in the below figure, let us determine V_x using the mesh current method.

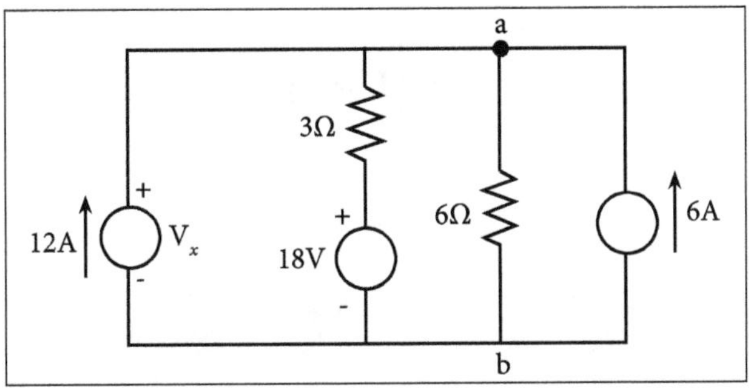

Solution:

Given:

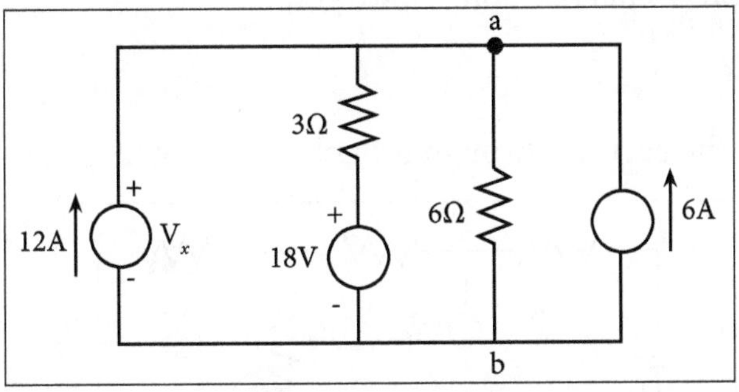

Convert the extreme right current source (6A) into a voltage source. It is important to note that the current source magnitude is 6A and its internal resistance is 6Ω.

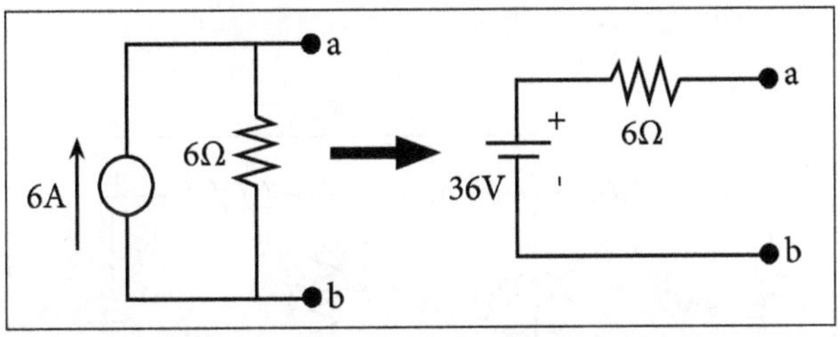

The given circuit is redrawn and is shown in the below figure.

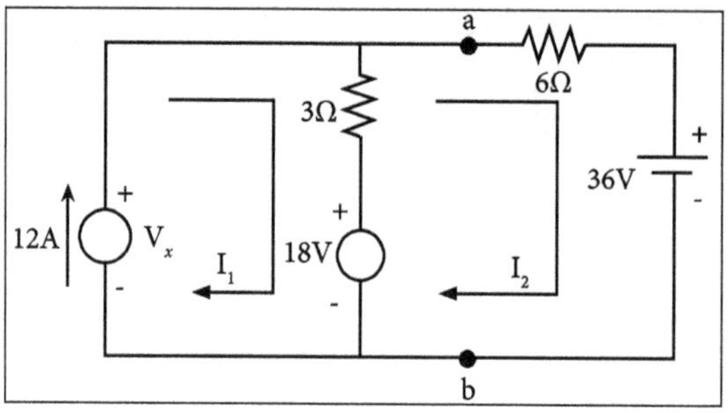

Loop 1:

By using KVL,

$$V_x - (I_1 - I_2) \times 3 - 18 = 0 \Rightarrow V_x + 3\,I_2 = 54 \qquad ...(1)$$

Where,

$$I_1 = 12\,A$$

Loop 2:

By using KVL,

$$18 - (I_2 - I_1) \times 3 - I_2 \times 6 - 36 = 0 \Rightarrow 9\,I_2 = 18 \Rightarrow I_2 = 2A \qquad ...(2)$$

Substituting equation (2) in equation (1), we get,

$$V_x = 48 \text{ volts.}$$

2. Let us determine the current through branch a-b using mesh analysis.

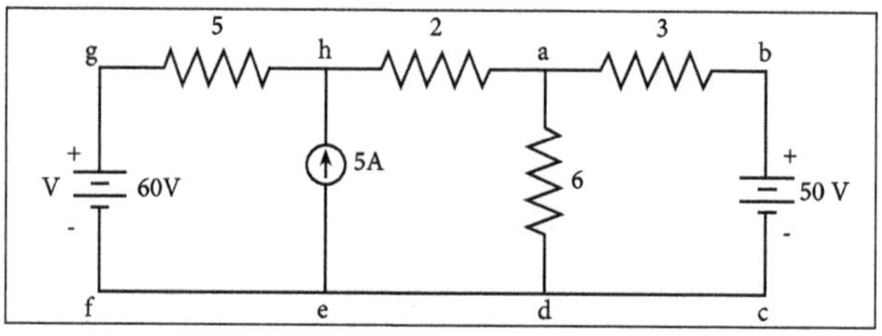

Solution:

Given:

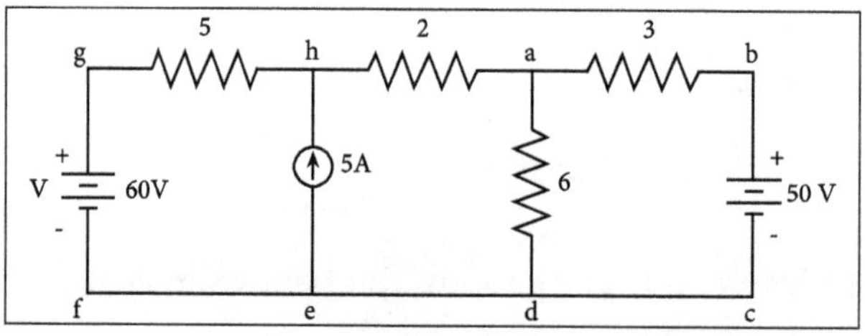

Consider the loops,

Loop HADEH,

$$5I_1 + 2I_2 + 6(I_2 - I_3) = 60$$

$$5I_1 + 8I_2 - 6I_3 = 60 \qquad ...(1)$$

Loop ABCDA,

$$3I_3 + 6(I_3 - I_2) = -50$$

$$3I_3 + 6I_3 - 6I_2 = -50$$

$$9I_3 - 6I_2 = -50 \qquad ...(2)$$

$$I_2 - I_1 = 5A \qquad ...(3)$$

From (1), (2) & (3),

$$D = \begin{vmatrix} -1 & 1 & 0 \\ 5 & 8 & -6 \\ 0 & -6 & 9 \end{vmatrix}$$

$$= -1(72 - 36) - 1(45)$$

$$D = -81$$

$$D_3 = \begin{vmatrix} -1 & 1 & 5 \\ 5 & 8 & 60 \\ 0 & -6 & -50 \end{vmatrix}$$

$$= -1(-400 + 360) - (-250) + 5(-30)$$

$$= 40 + 250 - 150$$

$$D_3 = 140$$

$$I_3 = D_3 / D$$

$=140/-81$

$=-1.7283$

The current through the branch ab is 1.7283 A.

3. Let us compute V_1 and V_2 in the circuit shown below, using nodal analysis.

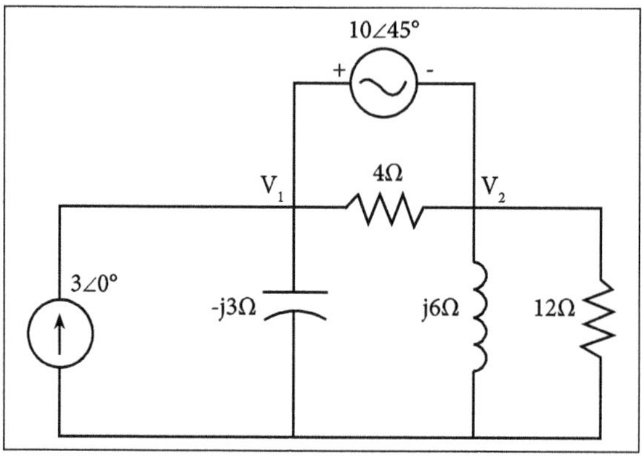

Solution:

Given:

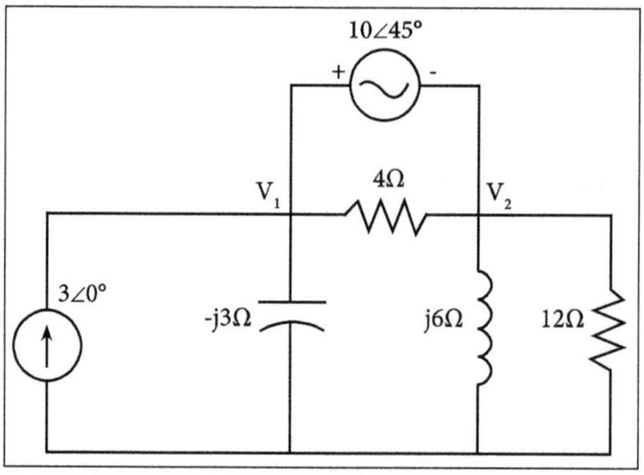

Step 1: Given voltage difference between V_1 and V_2 is,

$$V_1 - V_2 = 10\angle 45^\circ$$

$$V_1 - V_2 = 7.07 + j7.07 \quad ...(1)$$

Step 2: By applying KCL at nodes V_1 & V_2,

$$\frac{V_1}{-j3}+\frac{V_2}{j6}+\frac{V_2}{12}=3\angle 0^\circ$$

$$V_1(j0.33)+V_2\left(\frac{1}{j6}+\frac{1}{12}\right)=3\angle 0°$$

$$V_1(j0.33) + V_2(0.0833-0.1666j)=3+j0 \qquad \ldots(2)$$

Step 3: From equation (1) and (2),

$$\begin{bmatrix} 1 & -1 \\ j0.333 & 0.0833-j0.1666 \end{bmatrix}\begin{bmatrix} V_1 \\ V_2 \end{bmatrix}=\begin{bmatrix} 7.07+j7.07 \\ 3 \end{bmatrix}$$

Step 4: To determine V_1 and V_2,

$$V_1=\frac{\Delta_1}{\Delta}=\frac{\begin{bmatrix} 7.07+j7.07 & -1 \\ 3 & 0.0833-j0.1666 \end{bmatrix}}{\begin{bmatrix} 1 & -1 \\ j0.333 & 0.0833-j0.1666 \end{bmatrix}}$$

$$=\frac{(0.0833-j0.1666)*(7.07+j7.07)+3}{(0.0833-j0.1666)+j0.333}$$

$$=\frac{4.7665-j0.5886}{0.0833+j0.1664}$$

$$= 8.636 - j24.322$$

$$V_1=\frac{\Delta_2}{\Delta}=\frac{\begin{bmatrix} 1 & 7.07+j7.07 \\ j0.333 & 3 \end{bmatrix}}{\begin{bmatrix} 1 & -1 \\ j0.333 & 0.0833-j0.1666 \end{bmatrix}}$$

$$=\frac{3-(7.07+j7.07)*(j0.333)}{(0.0833-j0.1666)+j0.333}$$

$$=\frac{5.354-j2.354}{0.0833+j0.1664}$$

$$= 1.567 - j31.390$$

The node voltages V_1 and V_2 are in polar form as,

$$V_1 = 8.636 - j24.322 \ (\text{or}) \ 25.80 \angle -70.45^\circ$$

$$V_2 = 1.567 - j31.390 \ (\text{or}) \ 31.429 \angle -87.14^\circ$$

4. Let us use mesh analysis to determine three mesh currents in the circuit shown below.

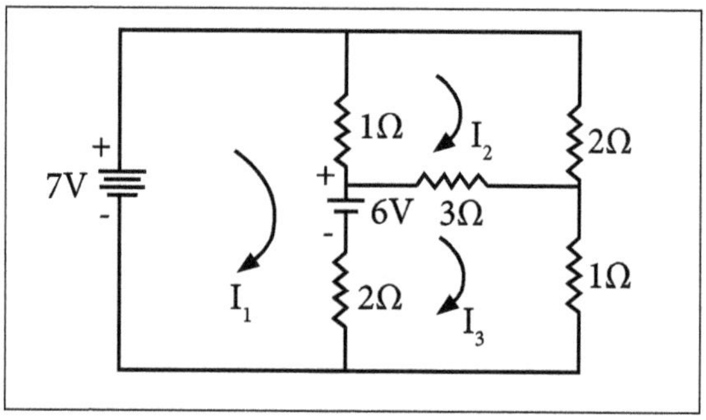

Solution:

Given:

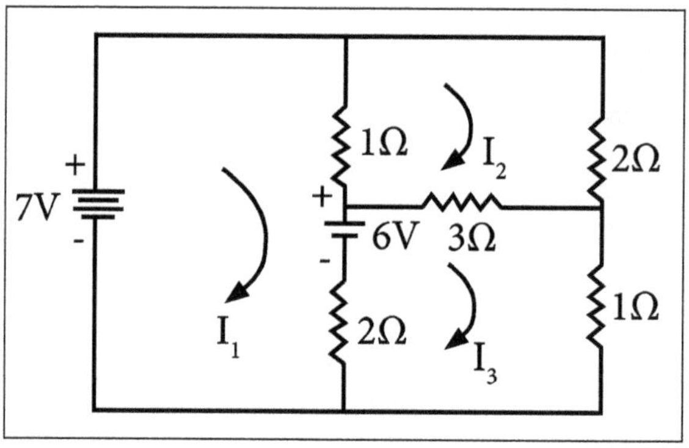

Mesh Loop 1:

$$-1(I_1 - I_2) - 6 - 2(I_1 - I_3) + 7 = 0$$

$$-3I_1 - I_2 - 2I_3 = 1 \qquad \text{...(1)}$$

Mesh Loop 2:

$$-2\,I_2 - 3(I_2 - I_3) - 1\,(I_2 - I_1) = 0$$

$$-I_1 + 6I_2 - 3I_3 = 0 \qquad \text{...(2)}$$

Mesh Loop 3:

$$I_3 - 2(I_3 - I_1) + 6 - 3(I_3 - I_2) = 0$$

$$-2I_1 - 3I_2 + 6I_3 = 0 \qquad \ldots(3)$$

Let us form the matrix from the above equations (1), (2) and (3),

$$\begin{bmatrix} 3 & -1 & -2 \\ -1 & 6 & -3 \\ -2 & -3 & 6 \end{bmatrix}\begin{bmatrix} I_1 \\ I_2 \\ I_3 \end{bmatrix} = \begin{bmatrix} 1 \\ 0 \\ 6 \end{bmatrix}$$

By applying Cramer's rule,

$$\Delta = \begin{bmatrix} 3 & -1 & -2 \\ -1 & 6 & -3 \\ -2 & -3 & 6 \end{bmatrix}$$

$$= 3(36-9) - (-1)(-6-6) - 2(3+12)$$

$$= 39$$

To find I_1:

$$\Delta_1 = \begin{bmatrix} 1 & -1 & -2 \\ 0 & 6 & -3 \\ 6 & -3 & 6 \end{bmatrix} = 1(36-9) - (-1)(18) - 2(-36)$$

$$= 117$$

$$I_1 = \frac{\Delta_1}{\Delta} = \frac{117}{39} \Rightarrow 3A$$

To find I_2:

$$\Delta_2 = \begin{bmatrix} 3 & 1 & -2 \\ -1 & 0 & -3 \\ -2 & 6 & 6 \end{bmatrix} = 3(18) - 1(-6-6) - 2(-6)$$

$$= 78$$

$$I_2 = \frac{\Delta_2}{\Delta} = \frac{78}{39} \Rightarrow 2A$$

To find I_3:

$$\Delta_3 = \begin{bmatrix} 3 & -1 & 1 \\ -1 & 6 & 0 \\ -2 & -3 & 6 \end{bmatrix} = 3(36) - (-1)(-6) + 1(3+12)$$

$$= 117$$

$$I_3 = \frac{\Delta_3}{\Delta} = \frac{117}{39} \Rightarrow 3A$$

5. Let us determine the voltage across 5Ω resistance and the current in the 12V source by using Nodal analysis.

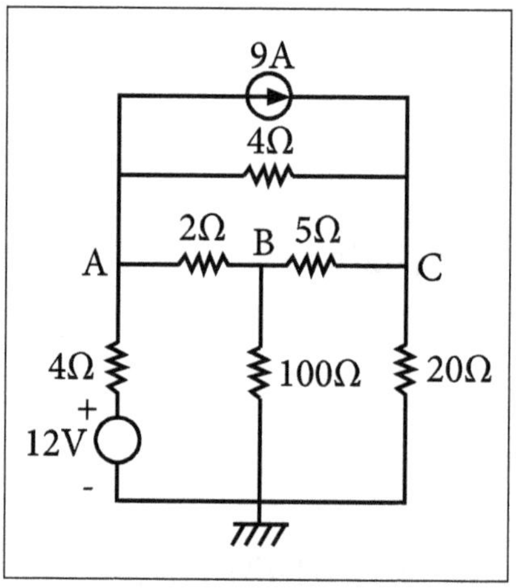

Solution:

Given:

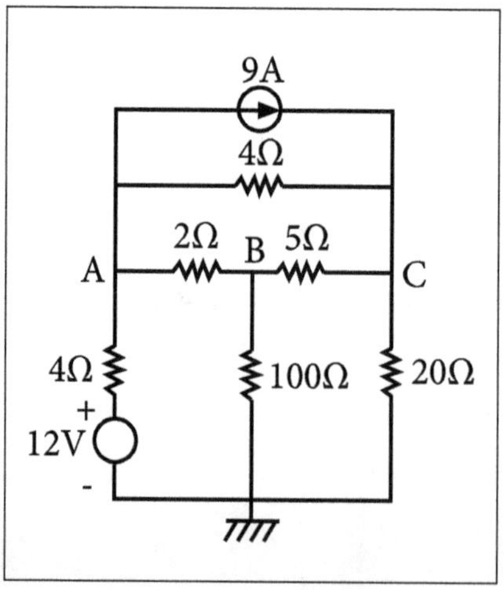

Using Nodal Method,

At Node A,

$$\frac{V_A - V_B}{2} + \frac{V_A - 12}{4} + \frac{V_A - V_C}{4} = -9$$

$$V_A\left(\frac{1}{2} + \frac{1}{4} + \frac{1}{4}\right) - \frac{V_B}{2} - \frac{V_C}{4} - \frac{12}{4} = -9$$

$$V_A - \frac{V_B}{2} - \frac{V_C}{4} = -6$$

$$4V_a - 2V_B - V_C = -\,24 \qquad \ldots(1)$$

At Node B,

$$\frac{V_B - V_A}{2} + \frac{V_B}{100} + \frac{V_B - V_C}{5} = 0$$

$$-V_A + V_B\left(\frac{1}{2} + \frac{1}{100} + \frac{1}{5}\right) - \frac{V_C}{5} = 0$$

$$-\frac{V_A}{2} + V_B\left(\frac{H}{100}\right) - \frac{V_C}{5} = 0$$

$$-50V_A + 71V_B - 20V_C = 0 \qquad \ldots(2)$$

At Node C,

$$\frac{V_C - V_B}{5} + \frac{V_C}{20} + \frac{V_C - V_A}{4} = 9$$

$$-\frac{V_A}{4} - \frac{V_B}{5} + V_C\left(\frac{1}{5} + \frac{1}{20} + \frac{1}{4}\right) = 9$$

$$-\frac{V_A}{4} - \frac{V_B}{5} + V_C\left(\frac{10}{20}\right) = 9$$

$$-5V_A - 4V_B + 10V_C = 180 \qquad \ldots(3)$$

To find V_A, V_B and V_C,

$$\Delta=\begin{vmatrix}4 & -2 & -1\\ -50 & 71 & -20\\ -5 & -4 & 10\end{vmatrix}=4(710-80)+2(-500-100)-1(200+355)$$

$$=2520-1200-555$$

$$=765$$

$$\Delta_B=\begin{vmatrix}4 & -24 & -1\\ -50 & 0 & -20\\ -5 & 180 & 10\end{vmatrix}=4(0+3600)+24(-500-100)-1(-9000+0)$$

$$=14400-14400+9000$$

$$=9000$$

$$V_B=\frac{\Delta_B}{\Delta}=\frac{9000}{765}=11.764\,\text{volts}$$

$$\Delta_C=\begin{vmatrix}4 & -2 & -24\\ -50 & 71 & 0\\ -5 & -4 & 180\end{vmatrix}=4(12780+0)+2(-9000+0)-24(200+355)$$

$$=51120-18000-13320=19800$$

$$V_C=\frac{\Delta_C}{\Delta}=\frac{19800}{765}=25.88\,\text{V}$$

∴ Current through 5Ω resistor is,

$$\frac{V_B-V_C}{5}=\frac{11.76-25.88}{5}=-14.12\,\text{Amps}$$

1.6 Types of Phasors, Sinusoidal Quantities, Relevant Theory and Principal of Duality

An alternating (ac) quantity (voltage, current or power) is defined as the one which changes its value as well as direction (polarity) with respect to time.

All our appliances such as TV, refrigerators, washing machines, air conditioners, fans etc., operate on the alternating voltage (ac voltage).

1.6.1 Definitions of Terms Associated with Periodic Functions

Time Period

Time period (T) is defined as the time taken in seconds by the waveform of an ac quantity to complete one cycle. After every T seconds, the cycle repeats itself as shown in the below figure.

Time period T = Time corresponding to one cycle.

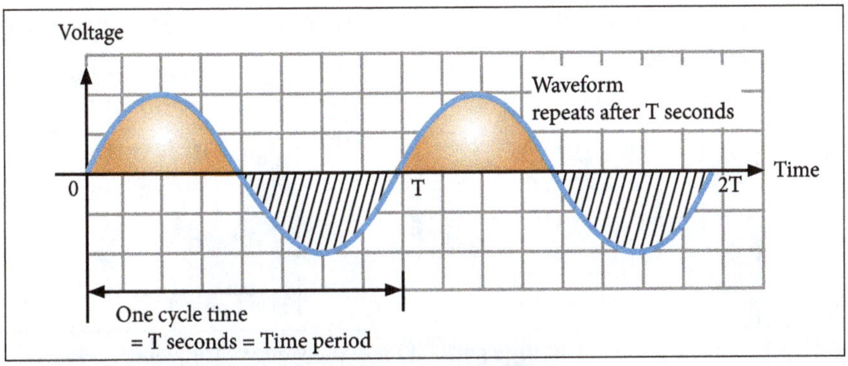

Concept of time period (T).

Angular Velocity

The angular velocity (ω) is the rate of change of angle ωt with respect to time,

$$\omega = \frac{d\theta}{dt}$$

Where,

$d\theta$ is the change in angle in time dt.

If dt = T i.e., time period, (one cycle) then the corresponding change in θ is 2π radians.

$$\therefore \ d\theta = 2\pi$$

$$\therefore \ \omega = \frac{2\pi}{T}$$

But,

$$1/T = f$$

$$\therefore \ \omega = 2\pi f$$

Frequency

Frequency is defined as the number of cycles completed by an alternating quantity in one second. It is denoted by "f" and its units are cycles/second or Hertz (Hz). As the time period (T) is the time in seconds per cycle denoted in seconds/cycle, hence relation between frequency and time period is as follows:

$$\text{Frequency}(f) = \frac{\text{cycles}}{\text{Second}} = \frac{1}{\text{Second / cycle}} = \frac{1}{T}$$

$$\therefore f = (1/T)\text{Hz.}$$

Therefore as the time period increases, the frequency decreases and vice-versa as shown in the below figure.

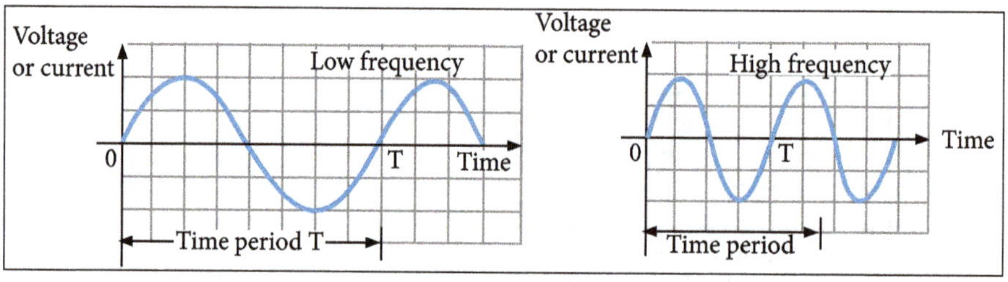

Effect of change in time period (T) on the value of frequency.

Thus, as we decrease the time period (T), its frequency increases as there are more number of cycles per second.

RMS Value

The effective or RMS value of an ac current is equal to the steady state or DC current that is required to produce the same amount of heat as produced by the ac current provided that the resistance and time for which these currents flow are identical.

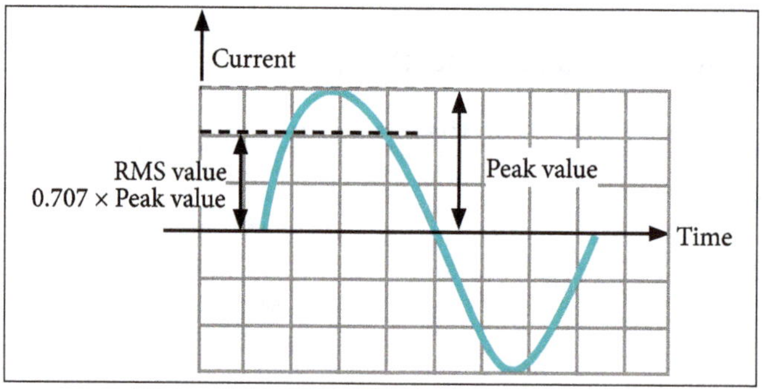

RMS value.

All the ac currents or voltages are expressed as RMS values unless clearly specified.

RMS value of ac current is denoted by I_{rms} and RMS voltage is denoted by V_{rms}. RMS value of a sinusoidal waveform (sine or cosine) is equal to 0.707 times its peak value.

$$I_{rms} = 0.707\ I_m$$

RMS value is called as the heat producing component of ac current. Amount of light produced by a lamp or the amount of heat produced by an iron is proportional to the square of rms value.

Average Value

The average value of an alternating quantity is equal to the average of all the instantaneous values over a period of half cycle.

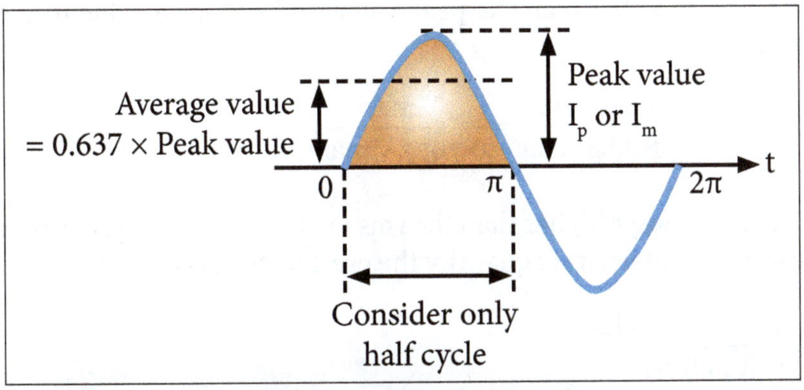

Average value of ac waveform.

The average value of ac current is denoted by I_{av} or I_{dc}. Average value is expressed without a positive or negative sign. The average value of a sinusoidal waveform is equal to 0.637 times its peak value.

$$I_{av} = I_{dc} = 0.637\ I_p = 0.637\ I_m$$

The dc ammeters or voltmeters indicate the average value. Average value of a full cycle of a symmetrical ac waveform is zero. The average value of a symmetrical alternating waveform is determined from just one half cycle of the waveform because the average value of the full cycle is zero.

Form Factor

The form factor of an alternating quantity is defined as the ratio of its RMS value to its average value.

$$\text{Form factor } K_f = \frac{\text{RMS value}}{\text{Average value}} \qquad ...(1)$$

Form factor is a dimensionless quantity and its value is always higher than one. That means the RMS value will always be higher than the average value. Form factor of a sinusoidal alternating current is given by,

$$K_f = \frac{I_{rms}}{I_{av}} = \frac{0.707\,I_m}{0.637\,I_m} = 1.11$$

The value of $K_f = 1.11$ is valid only for the sinusoidal (sine or cosine) ac quantities. For all the other shapes of ac quantities the form factor is different from 1.11 and it can be obtained by using the general definition given in equation (1).

Peak Factor or Crest Factor

Crest Factor is the ratio between the peak value and the RMS value of the waveform and is given as,

$$\text{Crest Factor} = \frac{\text{Peak value}}{\text{R.M.S value}} = \frac{\text{V max}}{0.707 \times \text{V max}}$$

As the peak value is always higher than the rms value, crest factor is always higher than one. For a sinusoidal alternating quantity the crest factor is given by,

$$K_p = \frac{\sqrt{2} \times \text{R.M.S value}}{\text{R.M.S value}}$$

$$K_p = \sqrt{2} \text{ or } 1.414$$

For a pure sinusoidal waveform the Crest Factor will always be equal to 1.414.

1.6.2 Phase Angle, Phasor Representation, Addition and Subtraction of Phasors

Phase is a measure of relative time difference between two sine waves. Even though phase is truly a time difference, it is always measured in terms of angle, either degrees or radians. This represents normalization to the time taken by one cycle of the wave in question, without regard to its true time period.

Phase Angle

The equation of the induced emf in the conductor is given by,

$$v = V_m \sin \omega t = V_m \sin \theta \qquad ...(1)$$

In the above equation (1), θ is the angle made by the conductor with the reference axis and it is called as the Phase Angle. Hence, the phase angle can take any value between 0 and 2π radians.

Phasor Representation

We can represent an alternating quantity such as a sine or cosine wave with the help of a phasor. So "phasor" representation which is a very simple way to represent a sinusoidal quantity is being used.

Definition and Graphical Representation of a Phasor

A phasor is a straight line with an arrow marked on one side. The length of this straight line represents the magnitude of the sinusoidal quantity being represented and the arrow represents its direction. Therefore, phasor representation is similar to the vector representation.

However the phasors rotate in the anticlockwise direction as shown in the below figure. The length of the phasor represents the rms value of the sinusoidal quantity. Sometimes the length also represents the peak value.

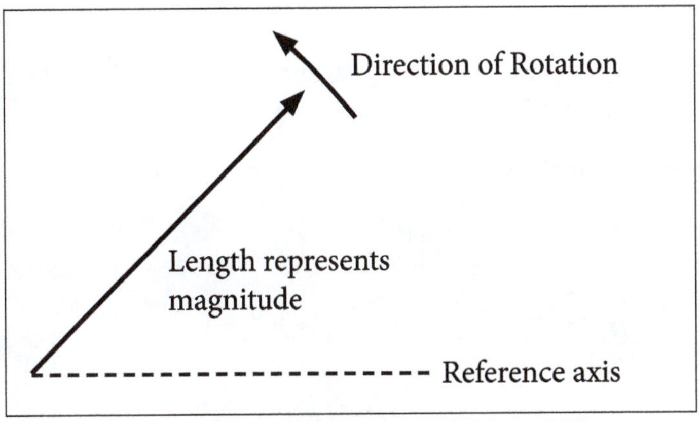

Phasor representation of sinusoidal quantities.

Speed of rotation of the phasor is equal to ω radians/sec, where $\omega = 2\pi f$. One rotation of the phasor corresponds to one cycle of the alternating waveform as shown in the below figure.

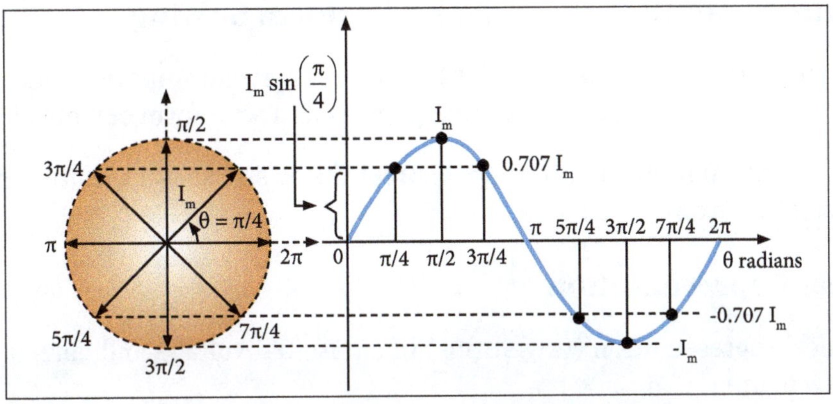

Relation between an alternating quantity and phasor.

If the length of the phasor is equal to the peak value of the sinusoidal ac quantity, its angular velocity $\omega = 2\pi f$ and it rotates in the anticlockwise direction in space then at any given angle θ, its projection on the Y-axis gives us the instantaneous value of the sinusoidal ac quantity at that angle.

We can represent two or more sinusoidal quantities simultaneously on the same phasor diagram if and only if their frequencies are same. If their frequencies are not same then they cannot be represented on the same phasor diagram.

Addition and Subtraction of Phasors

If two phasors are to be added or subtracted, then we have to represent them in the rectangular format.

Let,

$$v_1 = r_1 \angle\phi_1 \text{ and } v_2 = r_2 \angle\phi_2.$$

To obtain $(v_1 + v_2)$ or $(v_1 - v_2)$ we should represent v_1 and v_2 in their rectangular forms as follows,

$$v_1 = r_1 \cos\phi_1 + r_1 \sin\phi_1 = x_1 + jy_1 \text{ and } v_2 = r_2 \cos\phi_2 + r_2 \sin\phi_2 = x_2 + jy_2$$

Hence,

$$v_1 + v_2 = (x_1 + jy_1) + (x_2 + jy_2) = (x_1 + x_2) + j(y_1 + y_2)$$

Similarly,

$$v_1 - v_2 = (x_1 + jy_1) - (x_2 + jy_2) = (x_1 - x_2) + j(y_1 - y_2)$$

1.6.3 Mathematical Representation of Sinusoidal Quantities, Explanation with Relevant Theory, Problem Solving

In the graphical representation, we plot the values of the ac sinusoidal quantity (such as voltage or current) with respect to time. It gives us a waveform of sinusoidal wave.

In the mathematical representation, we express the ac sinusoidal quantity in the form of a mathematical expression.

Graphical Representation

The graphical representation (waveforms) of sinusoidal voltage and current is shown in the below figure.

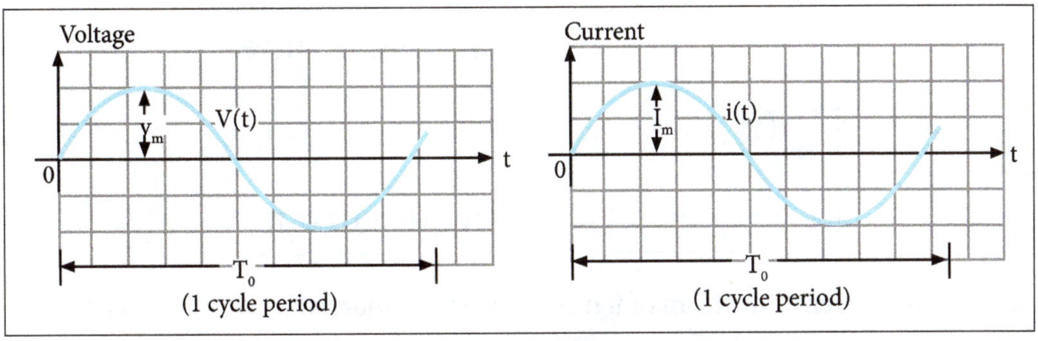

Instantaneous sinusoidal voltage and Instantaneous sinusoidal current.

For plotting such a waveform we plot time (t) on x-axis and corresponding voltage v (t) or current i (t) on the y-axis. That means the graphical representation of sinusoidal voltage or current shows the instantaneous variation in the voltage or current with respect to time.

Mathematical Representation

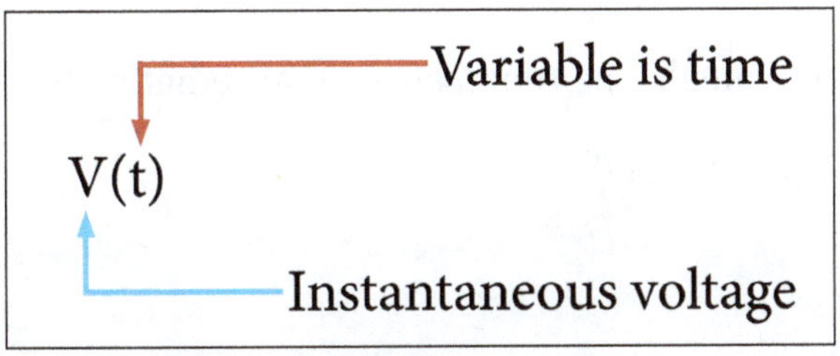

The voltage waveform of figure (a) is mathematically represented as,

$$v(t) = V_m \sin(2\pi f_0 t)$$

Where,

$v(t)$ = Instantaneous voltage,

V_m = Peak value (or maximum value)

f_0 = Frequency in Hz $(f_0 = 1/T_0)$

And "sin" represents the shape of the waveform.

It can also be represented as,

$$v(t) = V_m \sin(\omega_0 t) \text{ or } V_m \sin\theta$$

Where,

$$\theta = \omega_0 t = 2\pi f_0 t$$

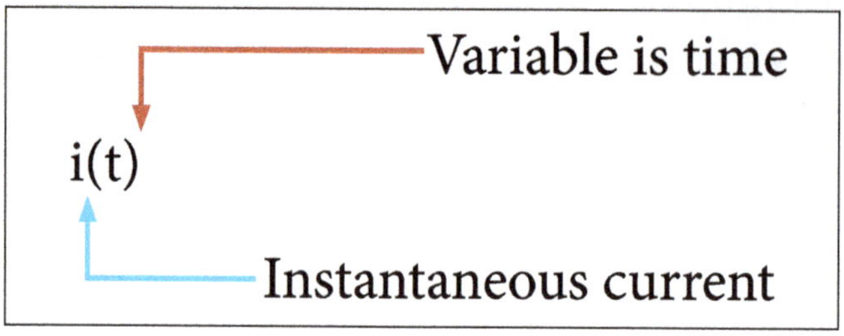

Similarly, the current waveform of figure (b) is mathematically represented as,

$$i(t) = I_m \sin(2\pi f_o t)$$

Where,

$i(t)$ = Instantaneous voltage

I_m = Peak value

f_o = Frequency in Hz

The alternating current can also be represented mathematically as,

$$i(t) = I_m \sin(\omega_o t) = I_m \sin\theta$$

Where,

$$\theta = \omega_o t = 2\pi f_o t$$

The sinusoidal waveform can be sine waves or cosine waves. The cosine voltage and current waveforms and their mathematical expressions are as shown in the below figure.

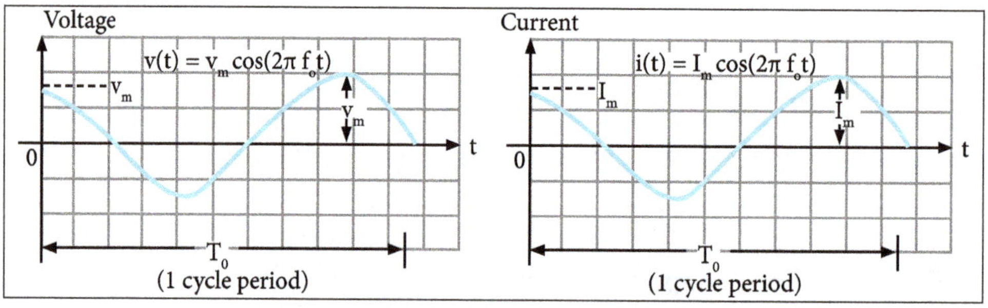

(a) Cosine voltage and its mathematical representation
(b) Cosine current and its mathematical representation.

Thevenins Theorem

In circuit theory, Thevenins theorem for linear electrical networks states that any combination of voltage sources, current sources and resistors with two terminals is electrically equivalent to the single voltage source V and the single series resistor R. For the single frequency AC systems, the theorem can also be applied to general impedances, not just for resistors.

Norton's Theorem

Norton's theorem states the following,

Any two-terminal linear bilateral dc network can be replaced by an equivalent circuit consisting of a current and a parallel resistor.

Problems

1. Let us determine the current in the 0.8 Ω resistor using Thevenin's theorem.

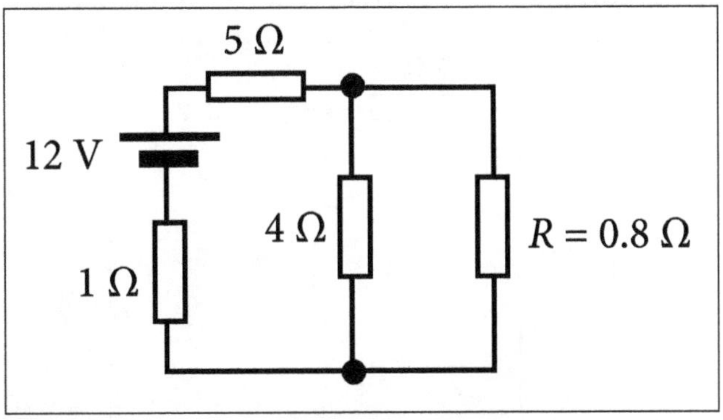

Solution:

Given:

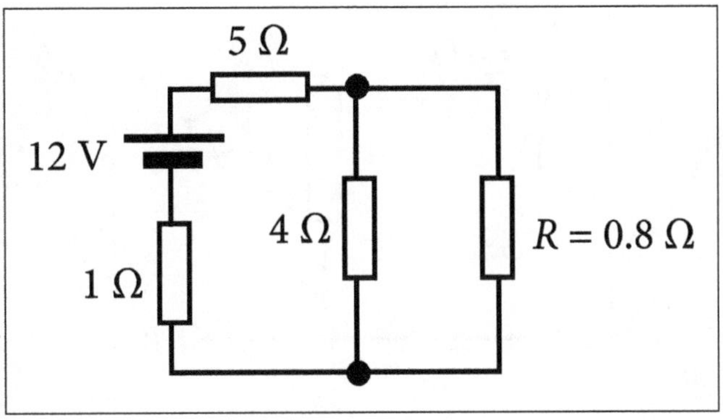

Following the procedure:

The 0.8 Ω resistors is removed from the circuit as shown in the figure.

Current $I_1 = 12/1+5+4 = 12/10 = 1.2A$

P.D. across 4 Ω resistor $= 4I_1 = (4)\ (1.2) = 4.8V$

Hence P.D. across AB, i.e., the open-circuit voltage across AB, E = 4.8V

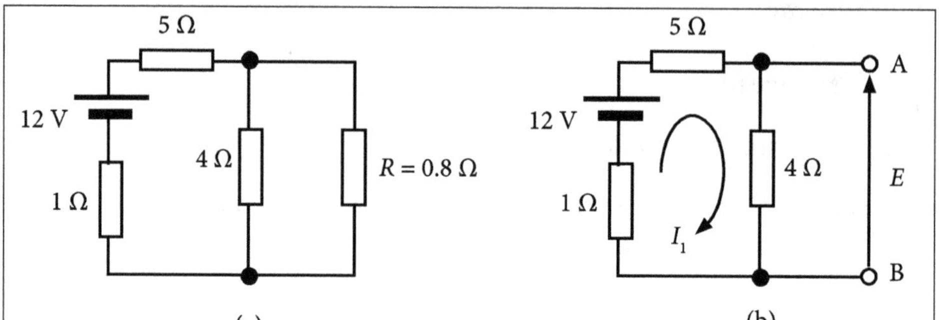

(a) (b)

Removing the source of e.m.f. gives the circuit shown in the below figure (c).

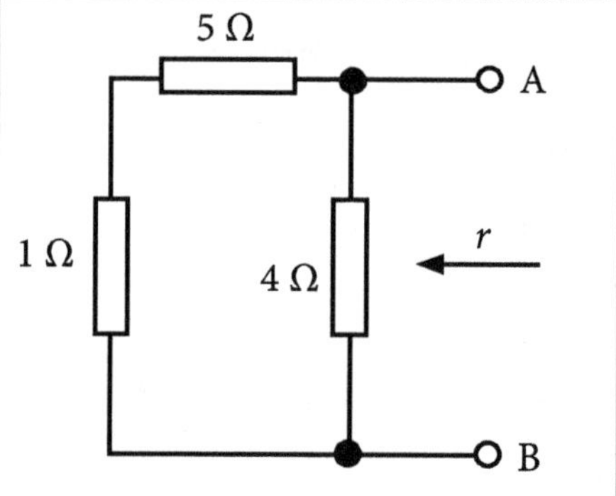

(c)

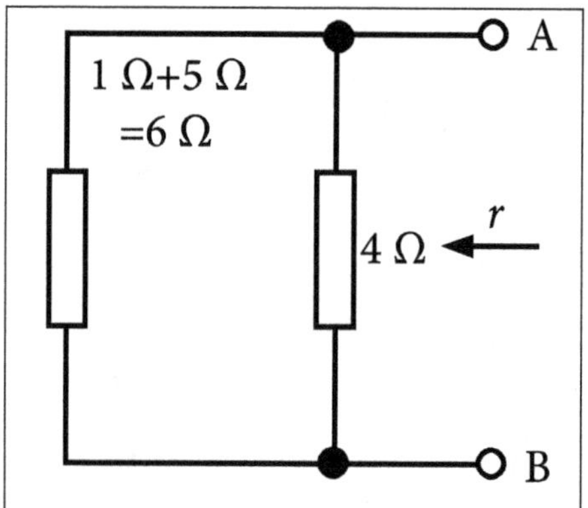

(d)

The equivalent circuit of Figure (c) is shown in above figure (d), from which the resistance,

$r = 4 \times 6 / 4 + 6 = 24 / 10$

$= 2.4\ \Omega$

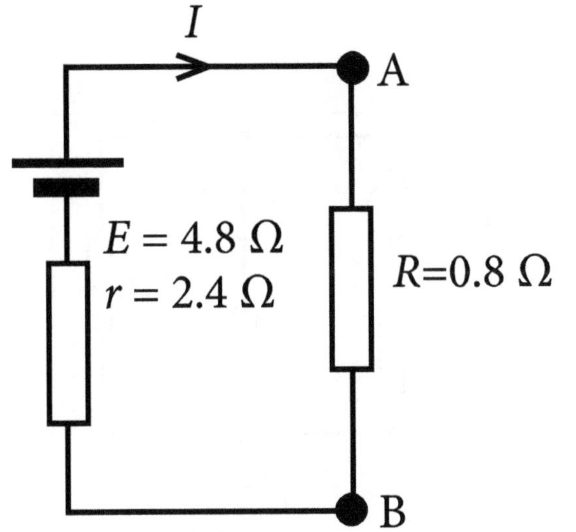

(e)

The equivalent Thevenin's circuit is shown in above figure (e), from which the current;

$I = E / r + R$

$= 4.8 / 2.4 + 0.8$

$= 4.8 / 3.2$

$I = 1.5A$ = current in the 0.8 Ω resistor

2. Use Norton's theorem to determine the current I flowing in the 4 Ω resistance shown in the below figure (a).

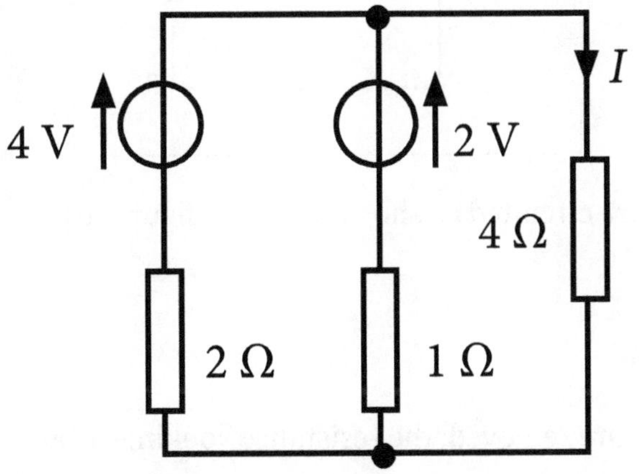

(a)

Solution:

Given:

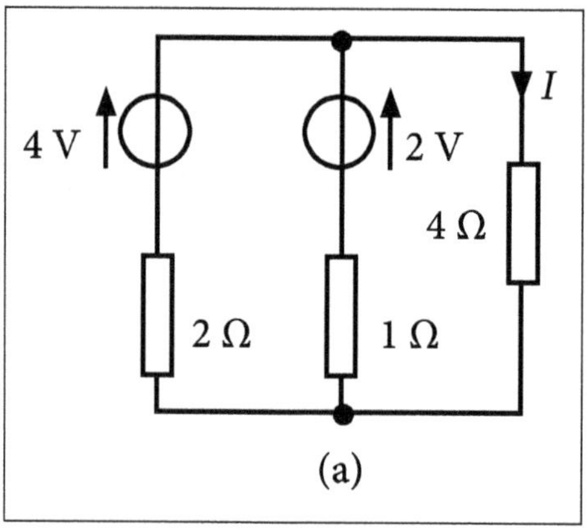

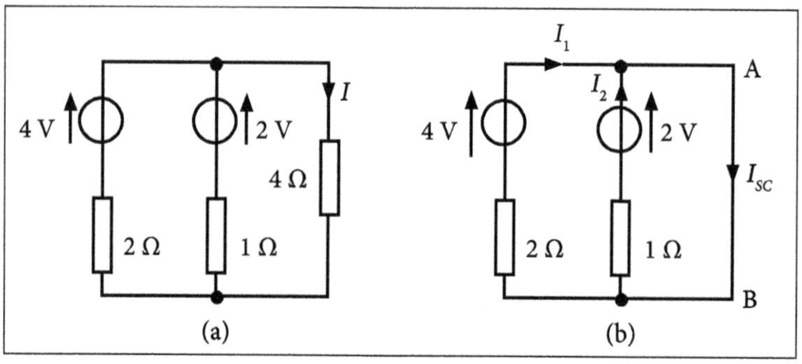

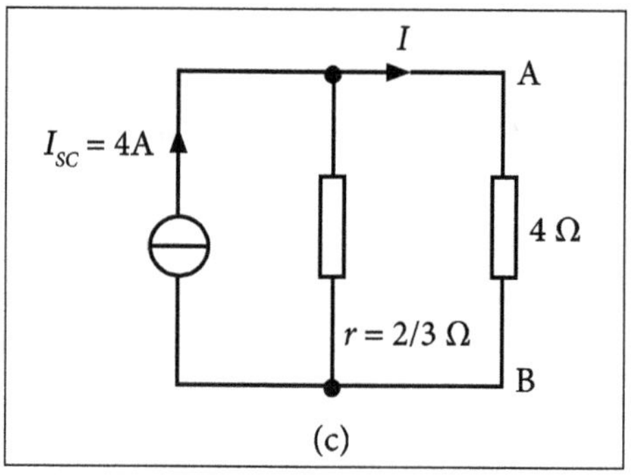

The 4 Ω branch is short-circuited as shown in above figure (b).

From Figure (b),

$$I_{SC} = I_1 + I_2 = 4A$$

If the sources of e.m.f. are removed, the resistance 'looking-in' at a break made between A and B is given by,

$$r = 2 \times 1/2 + 1 = 2/3\ \Omega$$

From the Norton equivalent network shown in Figure (c) the current in the 4 Ω resistance is given by,

$$I=(2/3/(2/3)+4)(4)=0.571A.$$

1.6.4 Principal of Duality with Examples

Duality is the mutual relationship. There is a number of similarities in analyzing the network on current (Loop) basis and voltage (Node) basis. The principal quantities involved in the two methods form pairs. Each of the quantity in such a pair thus plays a dual role. These quantities forming pair are called dual quantities.

Let us consider a network containing R, L & C elements connected in series and excited by a voltage source as shown in the below figure.

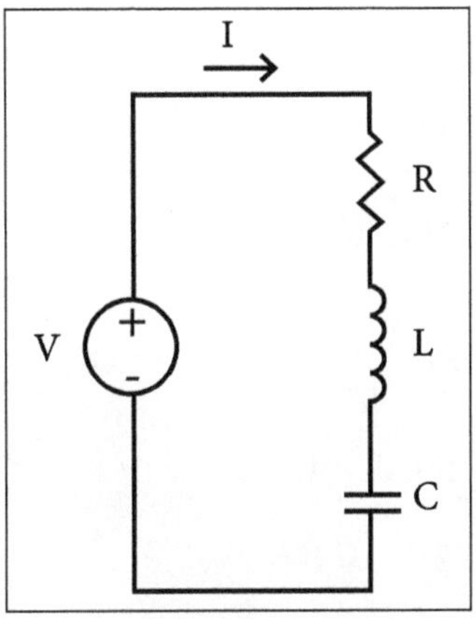

(a) Series RLC circuit.

The integral differential equations for the circuit is given by,

$$Ri+L(di/dt)+(1/C)\int i\, dt=v \qquad ...(1)$$

Let us consider a network containing R, L & C elements connected in parallel and driven by a current source as shown in the below figure.

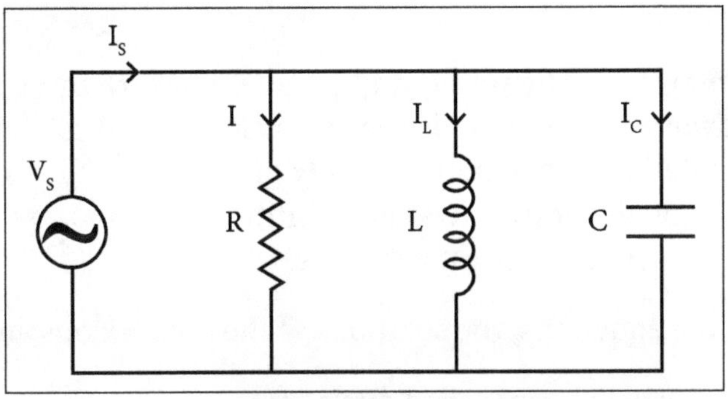

(b) Parallel RLC circuit.

The integral differential equations for the circuit is given as:

$$(1/R)\,v + C\,(dv/dt) + (1/L)\int v\,dt = i \qquad ...(2)$$

or,

$$G\,v + C(dv/dt) + (1/L)\int v\,dt = i$$

The equations governing the circuit behaviour are given by,

(i) $L_a \frac{di_a}{dt} + R_a i_a + \frac{1}{C_a}\int i_a dt = \upsilon_a$

$$\Rightarrow\ L_a \frac{di_a}{dt} + R_a i_a + \frac{1}{C_a} = \upsilon_a$$

$$\Rightarrow\ L_a \frac{di_a}{dt} + R_a i_a + \frac{1}{C_a}\int i_a dt = \upsilon_a$$

(ii) $C_b \frac{dv_b}{dt} + G_b \upsilon_b + \frac{1}{L_b}\int \upsilon_b dt = i_b$

$$\Rightarrow\ C_b \frac{d\upsilon_b}{dt} + G_b \upsilon_b + \frac{1}{L_b}\int \upsilon_b dt = i_b$$

$$\Rightarrow\ C_b \frac{d\upsilon_b}{dt} + G_b \upsilon_b + \frac{1}{L_b}\int \upsilon_b dt = i_b$$

Comparing the above two equations, we get the similarity between the networks of figure (a) and figure (b).

$$L_a \rightarrow C_b,\ R_a \rightarrow G_b,\ C_a \rightarrow L_b$$
$$\upsilon_a \rightarrow i_b,\ i_a \rightarrow \upsilon_b,$$

Hence, the series network in figure (a) and parallel network in figure(b) are duals of each other. The advantage of duality is that there is no need to analyze both types of circuits, since the solution of one automatically gives the solution of the other with a suitable change of symbols for the physical quantities. The below table gives the corresponding quantities for dual electrical networks.

If we observe both the equations, the solutions of these equations are the same.

Therefore, these two networks are called duals.

Construction of a Dual of a Network

Only planar networks without mutual inductances have duals.

S. No.	Quantity or Concept	Dual Quantity or Concept
1.	Current	Voltage
2.	Resistance	Conductance
3.	Inductance	Capacitance
4.	Branch current	Branch voltage
5.	Mesh	Node
6.	Loop	Node-pair
7.	Number of Loops	Number of Node-Pairs
8.	Loop Current	Node-Pair Voltage
9.	Mesh Current	Node Voltage or Node Potential
10.	Link	Tree Branch
11.	Tie-set	Cut-set
12.	Short Circuit	Open Circuit
13.	Parallel Path	Series Path
14.	Charge (Q)	Flux Linkages (ψ)

Procedure to Draw a Dual Network

Step 1: In each loop of a given network place a node and place an extra node called reference node outside the network.

Step 2: Draw the lines connecting adjacent nodes passing through each element and also to the reference node by placing the dual of each element in the line passing through original elements.

Table of Dual Quantities

S. No.	Loop basis	Node basis
1.	Voltage sources	Current sources

2.	A loop made up of several branches.	A node joining the same number of branches.
3.	Inductances	Capacitances
4.	Loop currents	Node voltages
5.	Capacitances	Inductances
6.	Resistances	Conductance

Only planar networks have duals. The duals of planar networks could be obtained by a graphical technique known as the dot method. The dot method has the following procedure:

- Put a dot in each independent loop of the network. This dot corresponds to the independent nodes in the dual network. Planar networks are those that can be laid on a plane without branches crossing one another.
- Put a dot outside the network. This dot corresponds to the reference node in the dual network.
- Connect all the internal dots in the neighboring loops by dashed lines cutting the common branches. These branches that are cut by the dashed lines will form the branches connecting the corresponding independent nodes in the dual network.
- As an example, if a common branch contains R and L in series, then the parallel combination of G and C should be put between the corresponding independent nodes in the dual network.
- Join all the internal dots to the external dot by using dashed lines cutting all the external branches. Duals of these branches cut by dashed lines will form the branches connecting the independent nodes and the reference node.
- Convention for sources in the dual network:
    - A voltage rise in the direction of a clockwise loop current corresponds to the current flowing toward the dual independent node.
    - A clockwise current source in the loop corresponds to the voltage source with a positive polarity at the dual independent node.

Problems

1. Let us draw the dual of the circuit shown in figure (a). Also let us write the mesh equations for the given network and node equations for its dual. Here we shall verify whether they are dual equations.

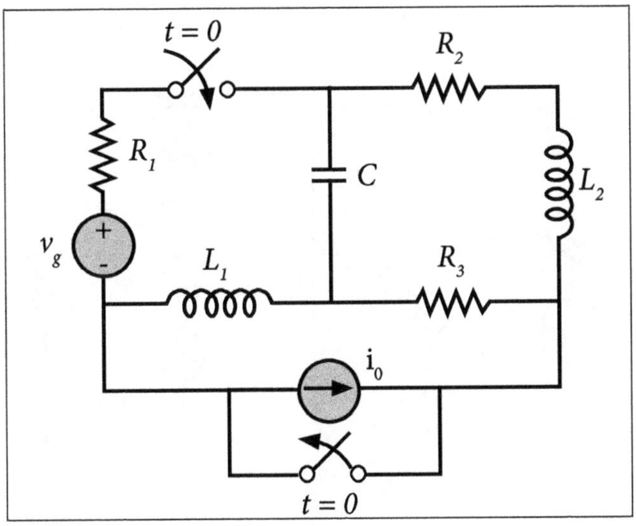

(a)

Solution:

Given:

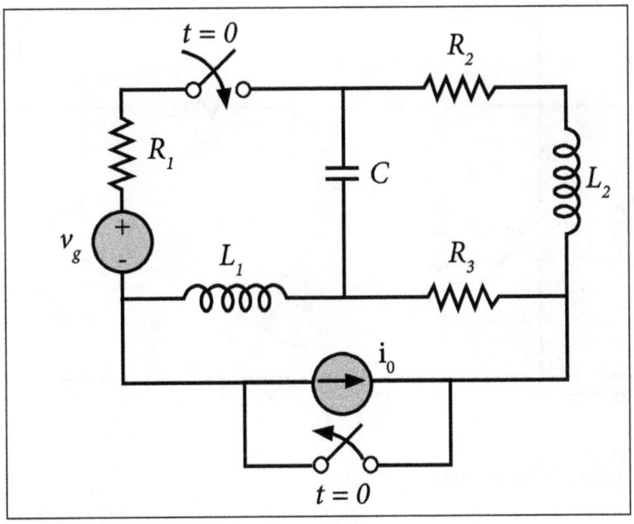

For the given network, the mesh equations are,

$$R_1 i_1 + L_1 D(i_1 - i_2) + \frac{1}{C}\int (i_1 - i_3)dt = \upsilon_g$$

$$i_2 = -i_o$$

$$R_2 i_2 + L_2 D i_3 + R_3(i_3 - i_2) + \frac{1}{C}\int (i_3 - i_2)dt = 0$$

The dual network, as per the procedure described in the theory is prepared as shown in figure(b) and is drawn as shown in figure(c).

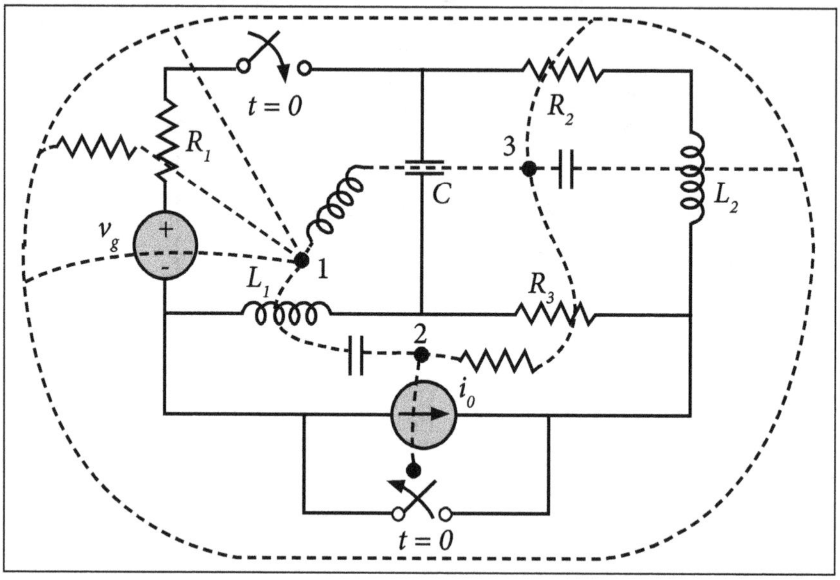

(b)

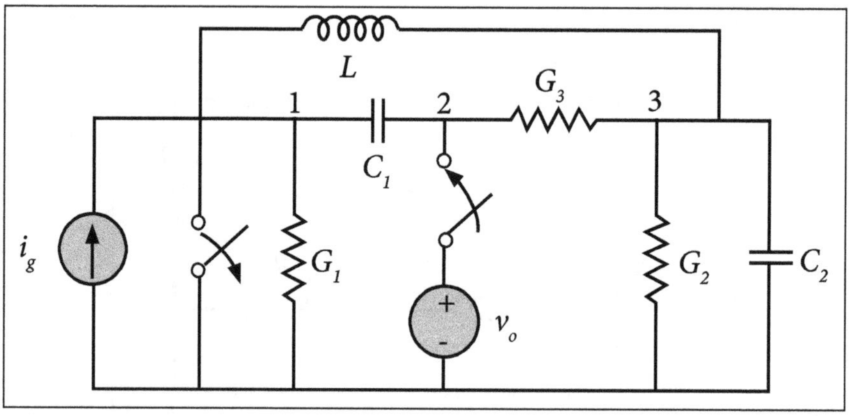

(c)

The node equations for this network are,

$$G_1V_1 + C_1D\left(\upsilon_1 - \upsilon_2\right) + \frac{1}{L}\int\left(\upsilon_1 - \upsilon_3\right)dt = i_g$$

$$G_2v_3 + C_2D\upsilon_3 + G_3\left(\upsilon_3 - \upsilon_2\right) + \frac{1}{L}\int\left(\upsilon_3 - \upsilon_2\right)dt = 0$$

2. For the bridge network shown in the figure (a), let us draw its dual and also let us write the integro-differential form of the mesh equations for the given network and node equations for its dual. The values for resistors are expressed in ohms, the capacitor values are expressed in farads and inductor values are expressed in Henrys.

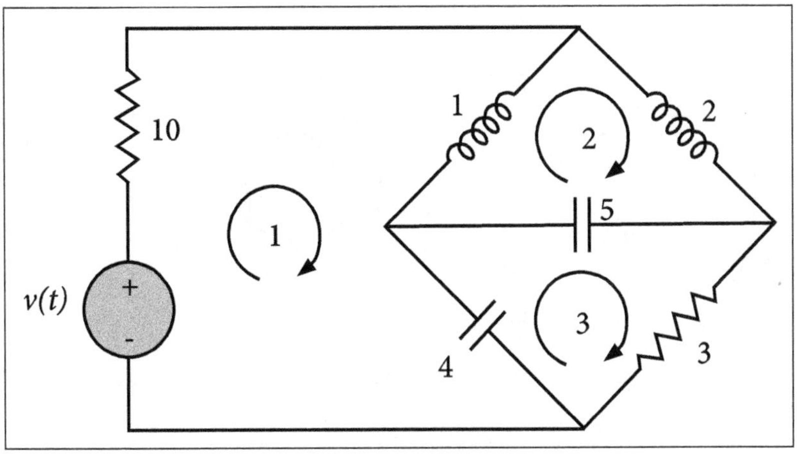

(a)

Solution:

Given:

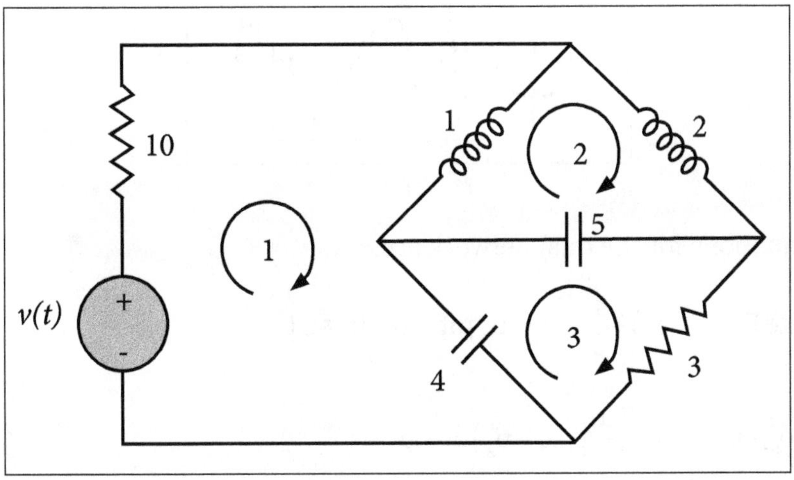

The dual for the given network is shown in figure(c) using the procedure shown in figure (b).

The integro-differential form for the network is,

$$10i_1 + D(i_1 - i_2) + \frac{1}{4}\int (i_1 - i_3)dt = 10\sin 50t$$

$$D(i_2 - i_1) + 2Di_2 \frac{1}{5}\int (i_2 - i_3)dt = 0$$

$$3i_3 + \frac{1}{4}\int (i_2 - i_3)dt + \frac{1}{5}\int (i_3 - i_2)dt = 0$$

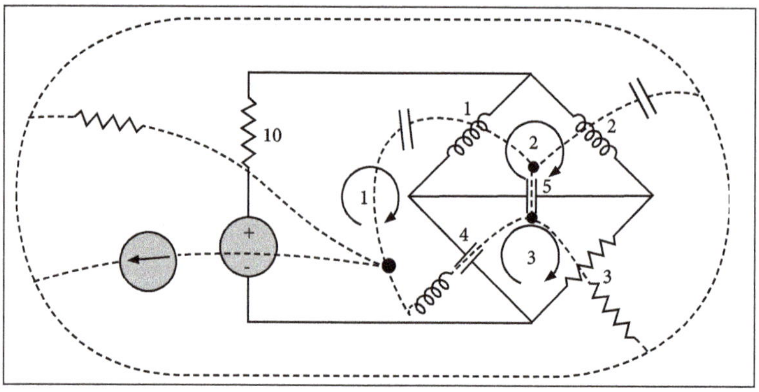

(b)

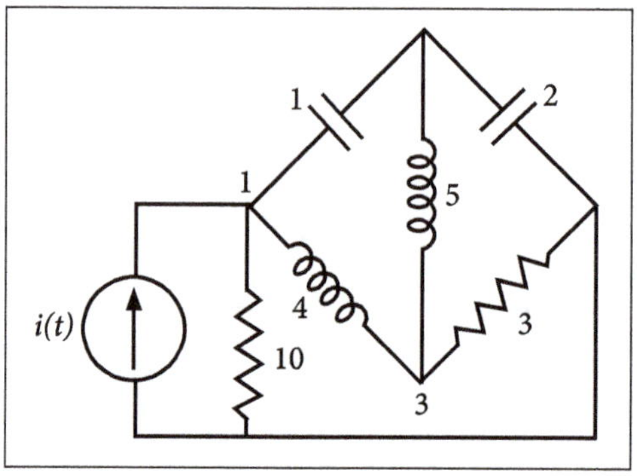

(c)

The node equations for the dual network are,

$$10\upsilon_1 + D(\upsilon_1 - \upsilon_2) + \frac{1}{4}\int(\upsilon_1 - \upsilon_3)dt = 10\sin 50t$$

$$D(\upsilon_2 - \upsilon_1) + 2D\upsilon_2 \frac{1}{5}\int(\upsilon_2 - \upsilon_3)dt = 0$$

$$3\upsilon_3 + \frac{1}{4}\int(\upsilon_3 - \upsilon_1)dt + \frac{1}{5}\int(\upsilon_3 - \upsilon_2)dt = 0$$

3. Let us draw the dual network of the one shown in figure below.

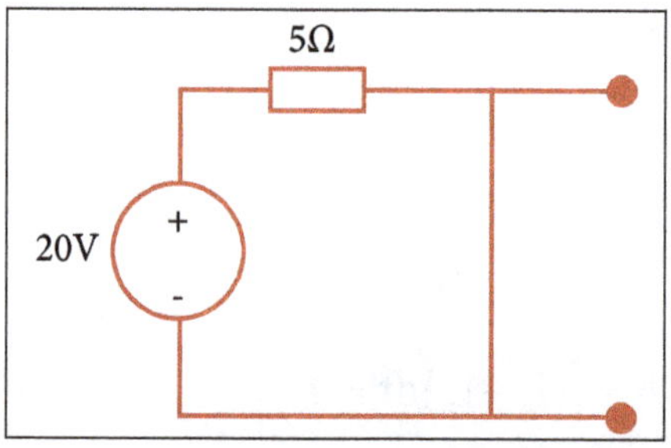

Solution:

Step 1: The dual of 20V voltage source is 20A current source, as shown in figure below.

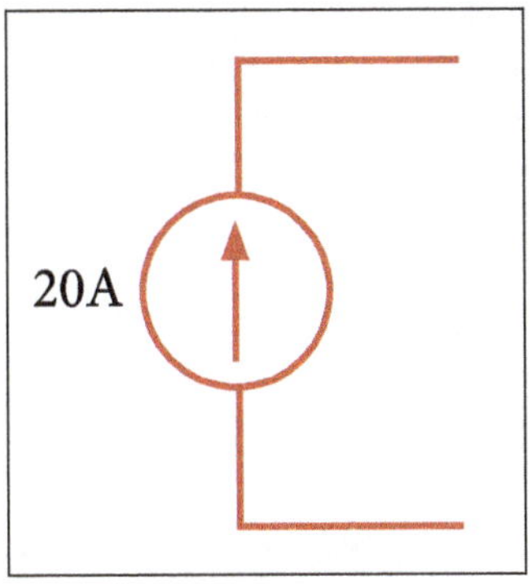

Step 2: The dual of 5Ω resistor is 1/5 series = 1/5Ω resistor in parallel.

Step 3: Since the 20V voltage source is in series with 5Ω resistor, it dual counterpart is 20A current source in parallel with 1/5Ω. This is shown in figure below.

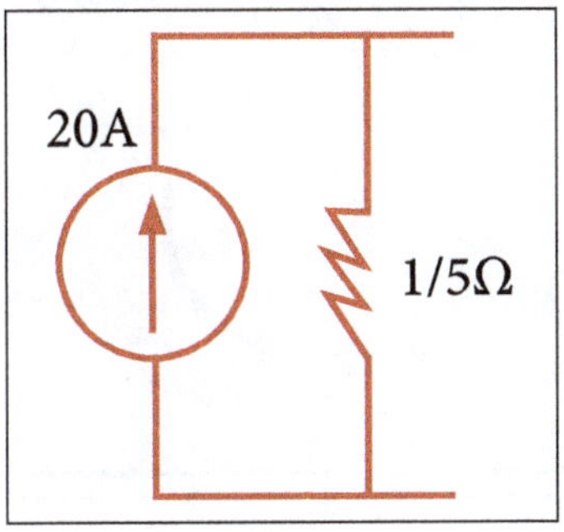

4. Let us draw the dual network of figure below.

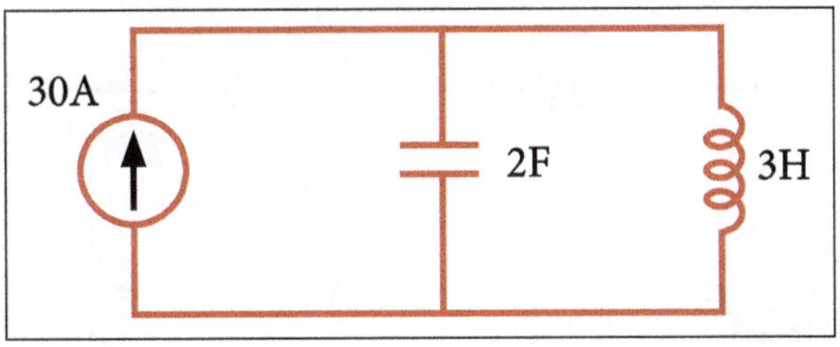

Solution:

Since the current source, capacitor and inductor are all connected in parallel, their dual counterparts will be connected in series.

- The dual of 30A current source is 30V voltage source.
- The dual of 2F capacitor is 2H inductor.

- The dual of 3H inductor is 3F capacitor.
- The dual network is obtained as,

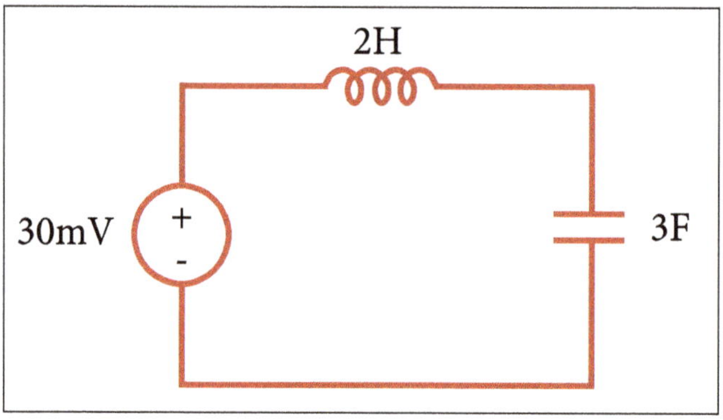

5. Let us draw the dual of a network for given network shown in the below figure.

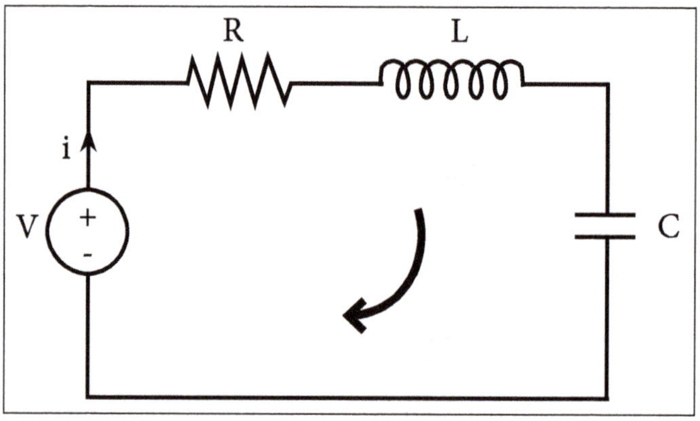

Solution:

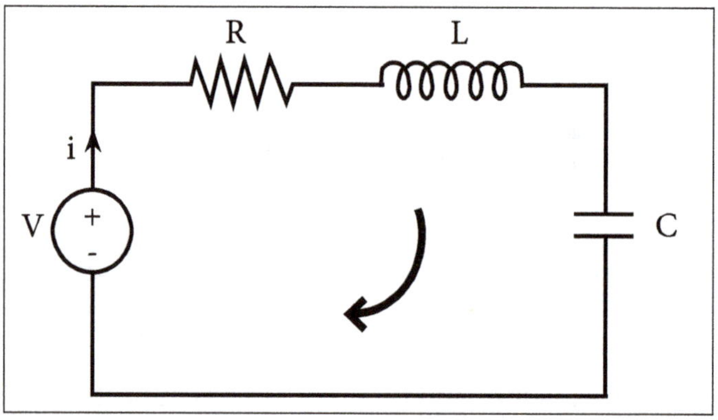

The integral differential equations for the circuit is given below as,

$$Ri + L\,(di/dt) + (1/C)\int i\,dt = v$$

Dual Network

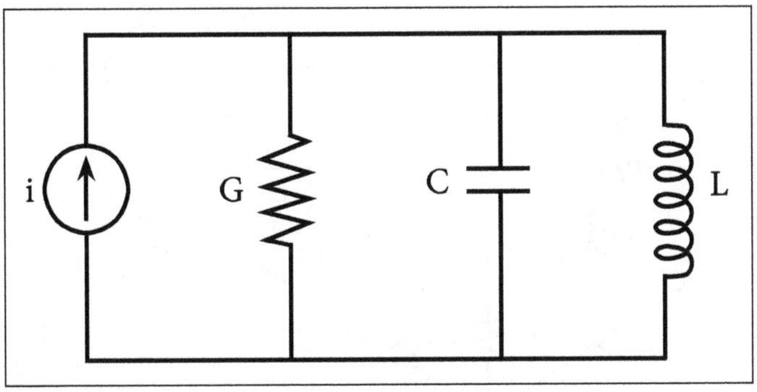

The integral differential equations for the circuit is given as,

$$(1/R)v + C\,(dv/dt) + (1/L)\int v\,dt = i$$

or,

$$Gv + C\,(dv/dt) + (1/L)\int v\,dt = i$$

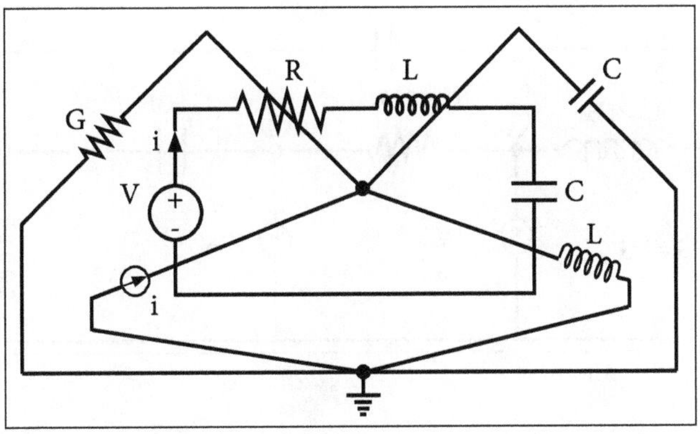

6. Let us draw the dual of the network shown in the below figure.

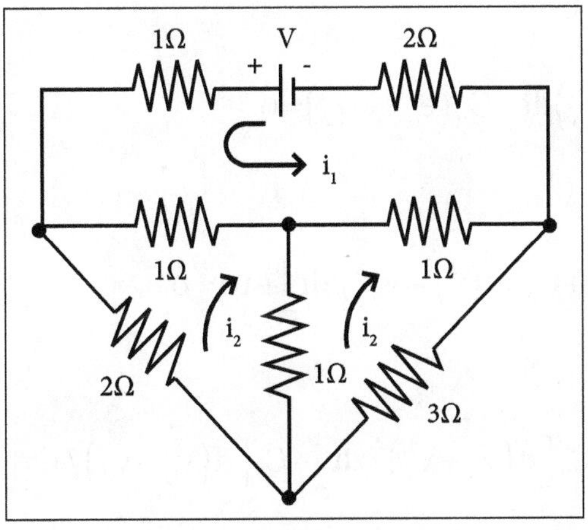

Solution:

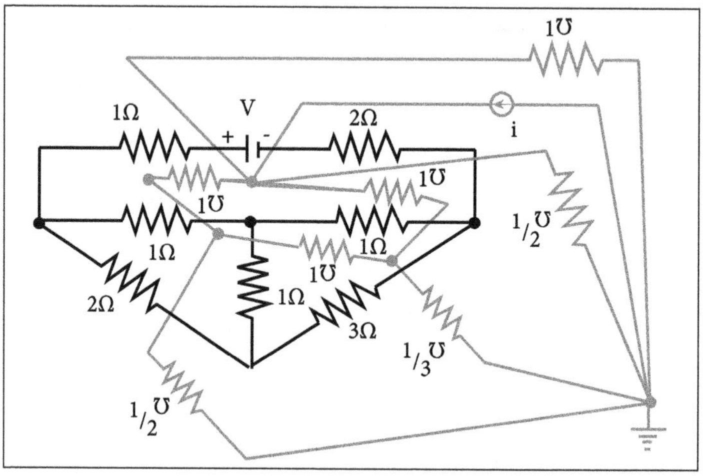

7. For the network shown in figure below, let us write the node equations. Let us draw the dual of this network and write mesh equations for the dual network and Verify whether these two sets of equations are dual equations.

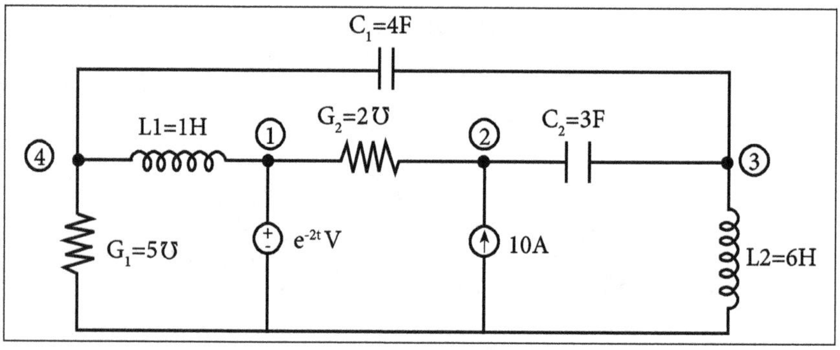

Solution:

Node equation:

At node (1),

$$(I/L_1)\int\left(e^{-2t}-v_4\right)dt+G_2\left(e^{-2t}-v_2\right)=0 \qquad ...(1)$$

At node (2),

$$(G_2)\int\left(v_2-e^{-2t}\right)+C_2\left[d\left(v_2-v_3\right)/dt\right]+10=0 \qquad ...(2)$$

At node (3),

$$(I/L_2)\int v_3\,dt+C_2\left[d\left(v_3-v_2\right)/dt\right]+C_1\left[d\left(v_3-v_4\right)/dt\right]=0 \qquad ...(3)$$

At node (4),

$$G_1 v_4 + (1/L_1)\int(v_4 - e^{-2t})dt + C_1[d(v_4 - v_3)/dt] = 0 \quad ...(4)$$

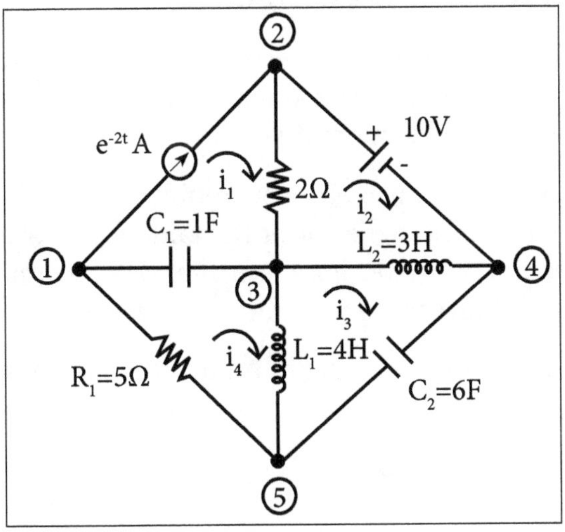

Loop equation for the dual network,

$$(1/C_1)\int(e^{-2t} - i_4)dt + R_2(e^{-2t} - i_4) = 0 \quad ...(5)$$

$$R_2\int(i_2 - e^{-2t}) + L_2[d(i_2 - i_3)/dt] + 10 = 0 \quad ...(6)$$

$$(1/C_2)\int i_3\,dt + L_2[d(i_3 - i_2)/dt] + L_1[d(i_3 - i_4)/dt] = 0 \quad ...(7)$$

$$R_1 i_4 + (1/C_2)\int(i_4 - e^{-2t})dt + L_1[d(i_4 - i_3)/dt] = 0 \quad ...(8)$$

By looking into two sets of equations, we can say that two sets of equations are dual equation.

1.7 Definitions of Terminology in Network Topology

Network topology is a graphical representation of electric circuits. It is useful for analyzing complex electric circuits by converting them into network graphs. Network topology is also called as Graph theory.

A network graph is a network in which all nodes and loops are retained but its branches are represented by lines. The voltage sources are replaced by short circuits and current sources are replaced by open circuits. (Sources without internal impedances or admittances can also be treated in the same way because they can be shifted to other branches by E-shift and/or I-shift operations.)

When all the elements in a network are replaced by lines with circles or dots at both ends, configuration is called the graph of the network.

Terminology used in Network Graph

- Path: A sequence of branches traversed in going from one node to another is called a path. A set of branches that may be traversed in an order without passing through the same node more than once.
- Node: A node point is defined as an end point of a line segment and exits at the junction between two branches or at the end of an isolated branch. It may be the interconnection of two or more branches or a terminal of a branch. Usually interconnections of three or more branches are nodes.
- Degree of a node: It is the number of branches incident to it.

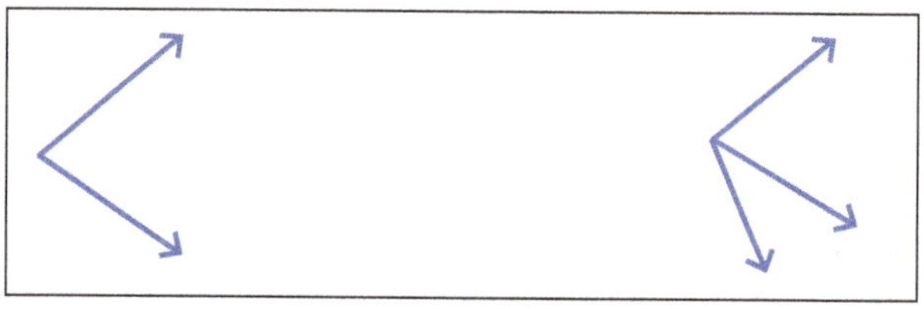

2-degree 3-degree.

- Tree: It is an interconnected open set of branches which include all the nodes of the given graph. In a tree of the graph there cannot be any closed loop.

A connected sub-graph containing all nodes of a graph but no closed path, i.e., it is a set of branches of graph which contains no loop but connects every node to every other node not necessarily directly. A number of different trees can be drawn for a given graph.

- Tree branch (Twig): It is the branch of a tree. It is also named as twig.
- Tree link (or chord): It is the branch of a graph that does not belong to the particular tree.
- Loop: This is the closed contour selected in a graph.
- Cut-Set: It is that set of elements or branches of a graph that separated two parts of a network. If any branch of the cut-set is not removed, the network remains connected. It is a set of branches that if removed divides a connected graph in to two connected sub-graphs.

Each cut set contains one tree branch and the remaining being links. The term cut-set is derived from the property designated by the way by which the network can be divided in to two parts.

- Tie-Set: It is a unique set with respect to a given tree at a connected graph containing on chord and all of the free branches contained in the free path formed between two vertices of the chord. A set of branches contained in a loop such that each loop contains one link and the remainder are tree branches.

- Network variables: A network consists of passive elements as well as sources of energy . In order to find out the response of the network the through current and voltages across each branch of the network are to be obtained.
- Directed (or Oriented) graph: A graph is said to be directed (or oriented) when all the nodes and branches are numbered or direction assigned to the branches by arrow.
- Sub graph: A graph said to be sub-graph of a graph G when its every node and branch is also a node and branch of G.
- Connected Graph: When at least one path along branches between every pair of a graph exits , it is called a connected graph.
- Incidence matrix: Any oriented graph can be described completely in a compact matrix form. Here we specify the orientation of each branch in the graph and the nodes at which this branch is incident. This branch is called incident matrix. When one row is completely deleted from the matrix the remaining matrix is called a reduced incidence matrix.
- Isomorphism: It is the property between two graphs so that both have got same incidence matrix.

Graph of a Network

It is collection of branches and nodes in which each branch connects two nodes. Also a network graph is a network in which all nodes and loops are retained but its branches are represented by lines. The voltage sources are replaced by short circuits and current sources are replaced by open circuits.

The diagram that gives network geometry and uses lines with dots at the ends to represent network element is usually called a graph of a given network. For example:

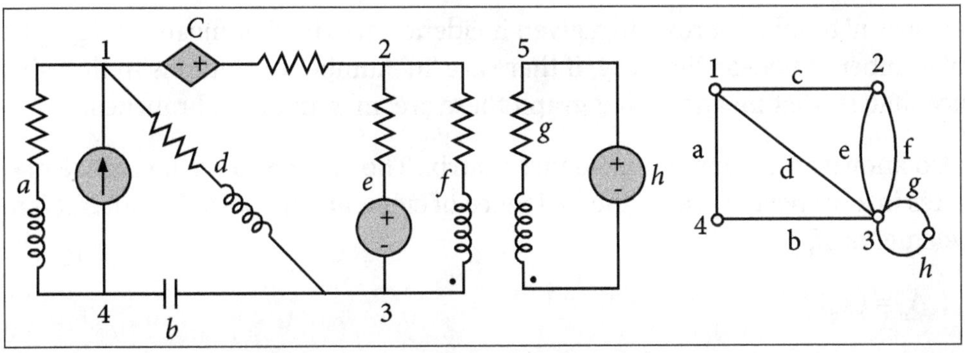

Network – Graph.

Planar Graph: A graph which may be drawn on a plane surface in such a way that no branch passes over any other branch.

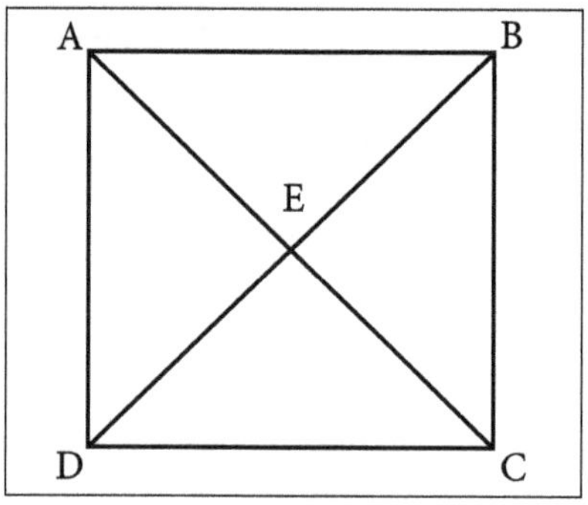

Planar graph.

Non-planar graph: Any graph which is not planar.

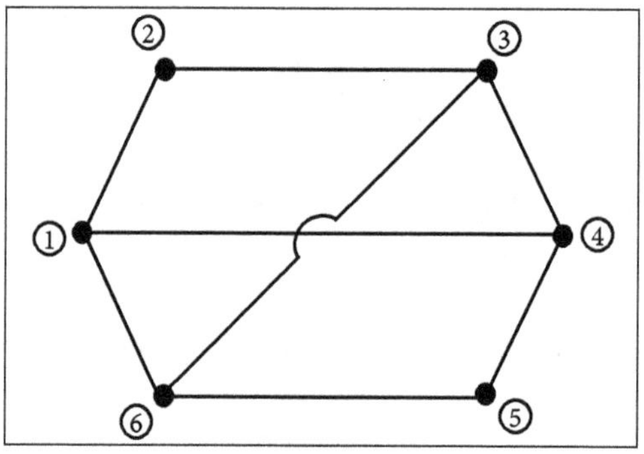

Non-planar graph.

Incidence Matrix

Incidence matrix is that matrix which represents the graph such that with the help of that matrix we can draw a graph. This matrix can be denoted as $[A_C]$. As in every matrix, there are also rows and columns in incidence matrix $[A_C]$. The rows of the matrix $[A_C]$ represents the number of nodes and the column of the matrix $[A_C]$ represents the number of branches in the given graph.

If there are 'n' number of rows in a given incidence matrix, that means in a graph there are 'n' number of nodes. Similarly, if there are 'm' number of columns in that given incidence matrix, that means in that graph there are 'm' number of branches.

It is also known as augmented incidence matrix. The element node incidence matrix A indicates in a connected graph, the incidence of elements to nodes. It is an N x B matrix with elements of,

$$A_n = (a_{kj})$$

$a_{kj} = 1$, when the branch b_j is incident to and oriented away from the k_{th} node.

= - 1, when the branch b_j, is incident to and oriented towards the k_{th} node.

= 0, when the branch b_j is not incident to the k_{th} node.

As each branch of the graph is incident to exactly two nodes,

$$\sum_{k=0}^{n} a_{kj} = 0 \quad \text{for} \quad j = 1,2,3...B.$$

That is, each column of An has exactly two non-zero elements, one being + 1 and the other - 1. Sum of elements of any column is zero. The columns of An are linearly dependent. The rank of the matrix is less than N.

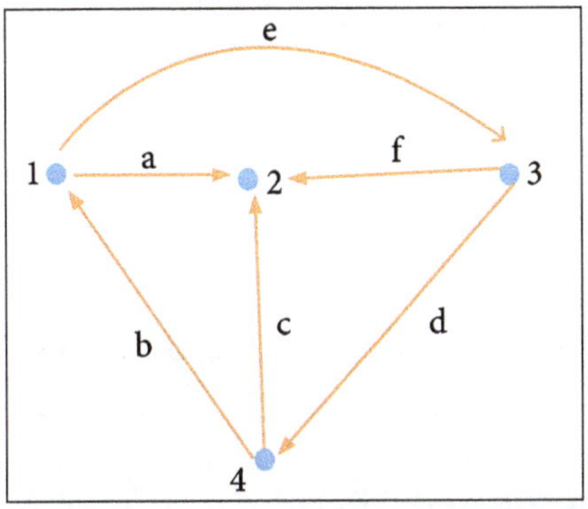

In the above shown graph or directed graph, there are 4 nodes and 6 branches. Thus the incidence matrix for the above graph will have 4 rows and 6 columns.

The entries of incidence matrix is always -1, 0, +1. This matrix is always analogous to KCL (Kirchhoff Current Law). Thus from KCL we can derive that,

Type of Branch Value

Outgoing branch from kth node +1.

Incoming branch to kth node -1.

Others 0.

Significance of the incidence matrix lies in the fact that it translates all the geometrical features in the graph into an algebraic expression.

By using the incidence matrix, we can write KCL as,

$A_n i_B = 0$,

Where,

i_B = branch current vector.

But these equations are not linearly independent. The rank of the matrix A is N - 1. This property of an is used to define another matrix called reduced incidence matrix or bus incidence matrix.

For the oriented graph shown in the below figure the incidence matrix is as follows,

Nodes ↓ branches

$$A_n = \begin{array}{c} \\ a \\ b \\ c \\ d \end{array} \begin{array}{c} \begin{array}{ccccc} 1 & 2 & 3 & 4 & 5 \end{array} \\ \begin{bmatrix} -1 & 1 & -1 & 0 & 0 \\ 1 & 0 & 0 & 1 & 0 \\ 0 & -1 & 0 & -1 & -1 \\ 0 & 0 & 1 & 0 & 1 \end{bmatrix} \end{array}$$

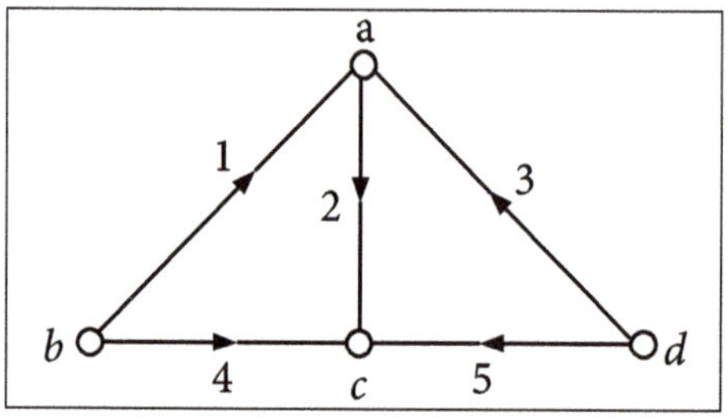

Steps to Construct Incidence Matrix

Following are the steps to draw the incidence matrix,

- If a given kth node has outgoing branch, then we will write +1.
- If a given kth node has incoming branch, then we will write -1.
- Rest other branches will be considered as 0.

Example of Incidence Matrix

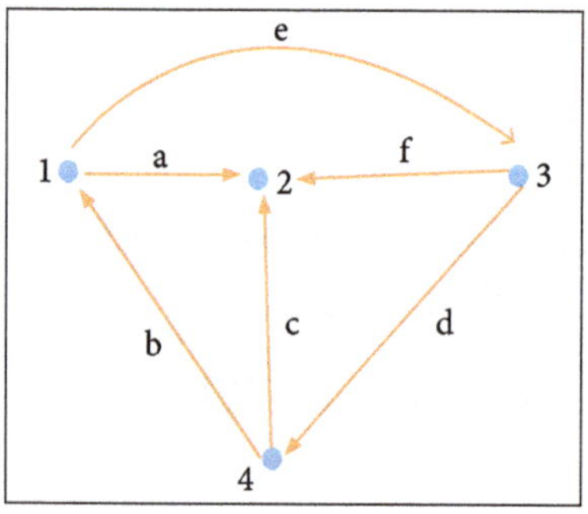

For the graph shown above, its incidence matrix is given as,

$[A_c] =$

nodes	branches	a	b	c	d	e	f
1		1	−1	0	0	1	0
2		−1	0	−1	0	0	−1
3		0	0	0	1	−1	1
4		0	1	1	−1	0	0

Points to Remember

- If sum of column comes to be zero, then the incidence matrix which we have created is correct else incorrect.
- For checking correctness of incidence matrix which we have drawn, we should check sum of column.
- The incidence matrix can be applied only to directed graph only.
- The rank of complete incidence matrix is (n-1), where n is the number of nodes of the graph.
- The number of entries in a row apart from zero tells us the number of branches linked to that node. This is also called as degree of that node.
- The order of incidence matrix is (n x b), where b is the number of branches of graph.
- From a given reduced incidence matrix, we can draw complete incidence matrix by simply adding either +1, 0 or -1 on the condition that sum of each column should be zero.

Tie Set Schedules: Twig Voltages and Cut Set Schedules

Tie-Set: A tie-set is a set of branches contained in a loop such that each loop contains one link or chord and the remainder are tree branches.

Let us consider the graph and the tree as shown below. This selected tree will result in three fundamental loops as we connect each link, in turn to the tree.

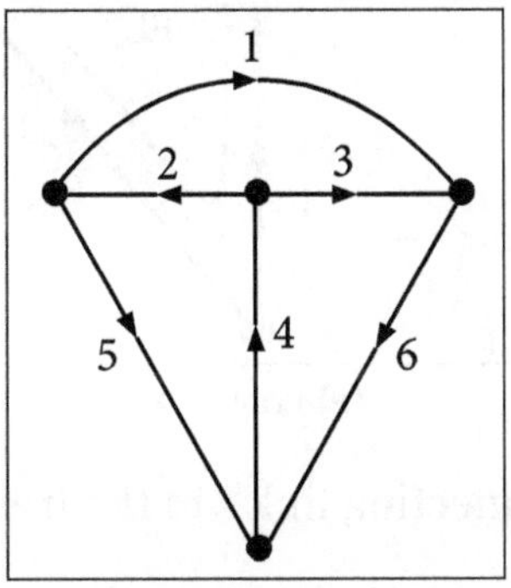

(a) Graph.

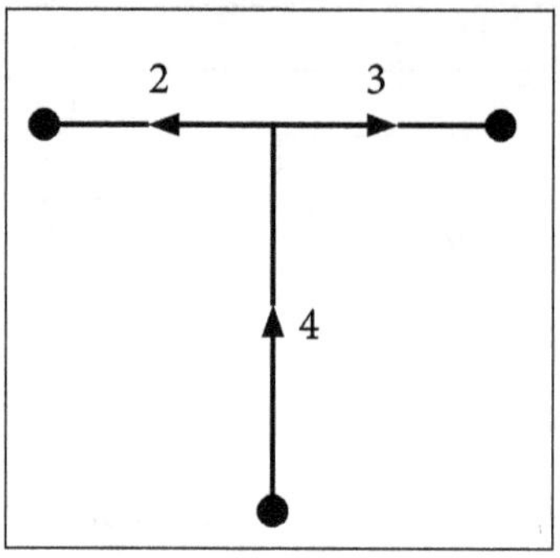

(b) Tree of the graph.

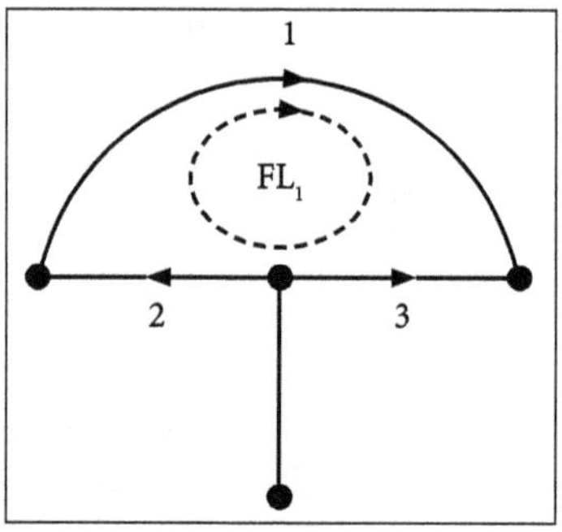

(c) Loop – 1.

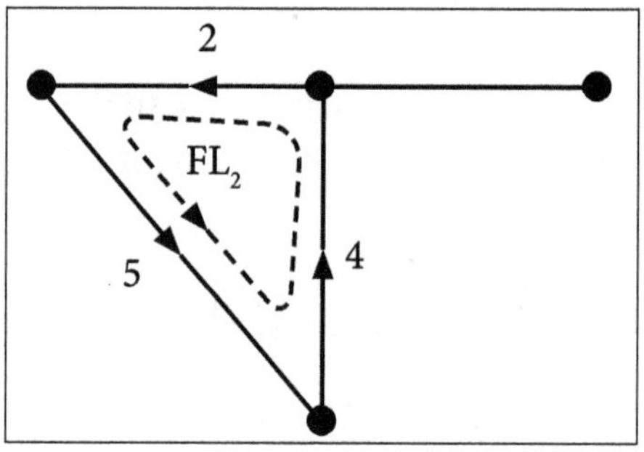

(d) Loop – 2.

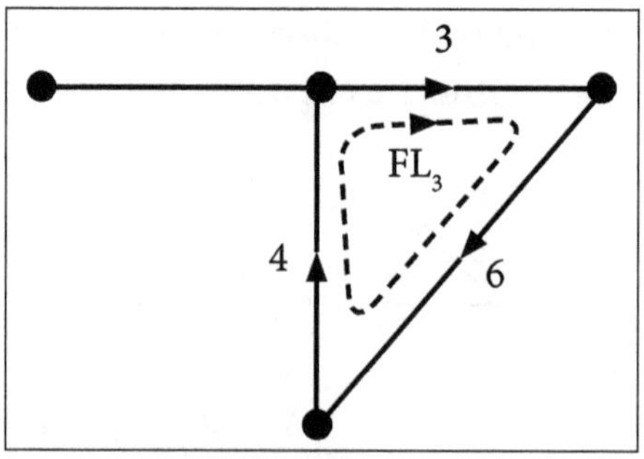

(e) Loop – 3.

Fundamental Loop 1 (FL_1): Connecting link 1 to the tree.

Fundamental Loop 2 (FL_2): Connecting link 5 to the tree.

Fundamental Loop 3 (FL_3): Connecting link 6 to the tree.

These sets of branches (1, 2, 3), (2, 4, 5) and (3, 4, 6) form three tie-sets.

For a given network tree, a systematic way of indicating the links is through use of a schedule called Tie-set Schedule.

The tie-set schedule for the given network oriented graph is shown below,

Link Current or Number	**Branches**							
	1	**2**	**3**	**4**	**5**	**6**	**7**	**8**
1	1	0	0	0	−1	0	0	1
2	0	1	0	0	1	−1	0	0
3	0	0	1	0	0	1	−1	0
4	0	0	0	1	0	0	1	−1

The tie-set schedule can be written in matrix form is known as Tie-set matrix (B).

$$B=\begin{pmatrix} 1 & 0 & 0 & 0 & -1 & 0 & 0 & 1 \\ 0 & 1 & 0 & 0 & 1 & -1 & 0 & 0 \\ 0 & 0 & 1 & 0 & 0 & 1 & -1 & 0 \\ 0 & 0 & 0 & 1 & 0 & 0 & 1 & -1 \end{pmatrix}$$

After writing the schedule, the columns of a schedule or matrix gives branch currents in terms of link currents. Thus, column 1 gives equation for j_1 in terms of link currents.

i.e.,

$$\left.\begin{array}{ll} j_1 = i_1 & j_5 = i_2 - i_1 \\ j_2 = i_2 & j_6 = i_3 - i_2 \\ j_3 = i_3 & j_7 = i_4 - i_3 \\ j_4 = i_4 & j_8 = i_1 - i_4 \end{array}\right\} \quad ...(1)$$

Similarly,

The rows of the schedule give KVL equations in terms of coefficients of the schedule or matrix.

$$\text{i. e.,} \left.\begin{array}{l} e_1 - e_5 + e_8 = 0 \\ e_2 + e_5 - e_6 = 0 \\ e_3 + e_6 - e_7 = 0 \\ e_4 + e_7 - e_8 = 0 \end{array}\right\} \quad ...(2)$$

The set of the above equation (1) can be expressed in the matrix form as follows,

$$
\begin{pmatrix} j_1 \\ j_2 \\ j_3 \\ j_4 \\ j_5 \\ j_6 \\ j_7 \\ j_8 \end{pmatrix} = \begin{pmatrix} 1 & 0 & 0 & 0 \\ 0 & 1 & 0 & 0 \\ 0 & 0 & 1 & 0 \\ 0 & 0 & 0 & 1 \\ -1 & -1 & 0 & 0 \\ 0 & 1 & -1 & 0 \\ 0 & 0 & 1 & 1 \\ 1 & 0 & 0 & -1 \end{pmatrix} \begin{pmatrix} i_1 \\ i_2 \\ i_3 \\ i_4 \end{pmatrix}
$$

...(3)

In compact form,

$$\left(I_b\right) = \left(B\right)^T \left(I_1\right)$$

...(4)

Where,

$\left(I_b\right)$ = is a column matrix of branch currents of the order (b × 1).

$\left(B\right)^T$ = is the transpose of the fundamental tie-set matrix B.

$\left(I_1\right)$ = is a column matrix of loop currents or link currents of the order (1 × 1).

Where,

'I' is the number of independent loops.

From the set of equation (2),

$$
\begin{pmatrix} 1 & 0 & 0 & 0 & -1 & 0 & 0 & 1 \\ 0 & 1 & 0 & 0 & 1 & -1 & 0 & 0 \\ 0 & 0 & 1 & 0 & 0 & 1 & -1 & 0 \\ 0 & 0 & 0 & 1 & 0 & 0 & 1 & -1 \end{pmatrix} \begin{pmatrix} e_1 \\ e_2 \\ e_3 \\ e_4 \\ e_5 \\ e_6 \\ e_7 \\ e_8 \end{pmatrix} = 0
$$

....(5)

In compact form,

$$\left(B\right)\left(E_b\right) = 0$$

...(6)

(E_b) = is a column matrix of branch voltages of the order (b × 1).

The above equations (4) and (6) are the basic equations required to deriving equilibrium equations on loop current basis.

Cut-set Schedule

For a given network tree, a systematic way of indicating the tree branch voltage through use of a schedule called cut-set schedule.

To write the cut-set schedule for network graph:

- Let us consider an oriented network graph.
- Write any one possible tree of the network graph.
- Let us assume tree branch voltages as $(e_1, e_2 ... e_n)$ independent variables.
- Let us also assume the independent voltage variable is same direction as that of a tree branch voltage.
- Mark the cut-sets (recognize) in the network graph.

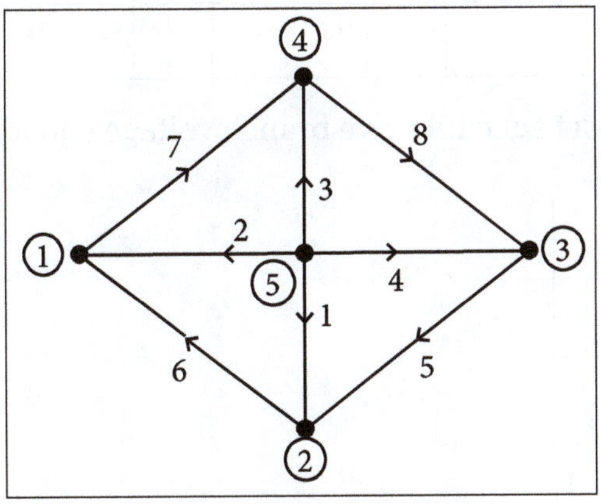

Oriented Graph.

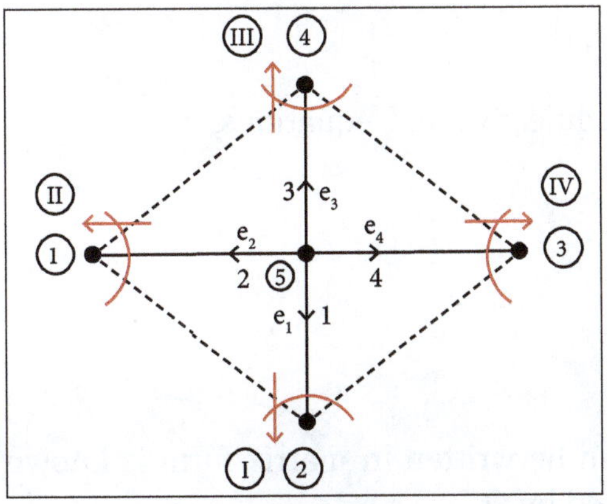

Tree and Cut-sets.

The tree branch voltages e_1, e_2, e_3, & e_4 entered in the first column of the schedule correspond to 4 branches 1, 2, 3 & 4. In order to fill the first row corresponding to the tree branch voltages e1, by looking into the direction of currents in the branches connected to the cut-set under consideration.

If the direction of current in the cut-set branch is towards the cut-set node, write '+1' in the branch column of concerned cut-set branch. If the direction of current in the cut-set branch is away from the cut-set node, write '-1' in that particular cut-set branch column. Write '0' in the branch columns, which are not in that particular, cut-set.

Cut-set Schedule:

Tree Branch Voltages	**Branches**							
	1	**2**	**3**	**4**	**5**	**6**	**7**	**8**
e_1	1	0	0	0	1	-1	0	0
e_2	0	1	0	0	0	1	-1	0
e_3	0	0	1	0	0	0	1	-1
e_4	0	0	0	1	-1	0	0	1

The columns of the cut-set schedule give branch voltage equations,

$$\text{i.e.,}\quad \left.\begin{aligned} v_1 &= e_1 \\ v_2 &= e_2 \\ v_3 &= e_3 \\ v_4 &= e_4 \\ v_5 &= (e_1 - e_4) \\ v_6 &= (-e_1 + e_2) \\ v_7 &= (-e_2 + e_3) \\ v_8 &= (-e_3 + e_4) \end{aligned}\right\} \quad \ldots(1)$$

The rows of cut-set schedule give KCL equations,

$$\left.\begin{aligned} j_1 + j_5 - j_6 &= 0 \\ j_2 + j_6 - j_7 &= 0 \\ j_3 + j_7 - j_8 &= 0 \\ j_4 - j_5 + j_8 &= 0 \end{aligned}\right\} \quad \ldots(2)$$

The cut-set schedule can be written in matrix form is known as cut-set matrix. This matrix can be represented by Q.

$$Q=\begin{pmatrix} 1 & 0 & 0 & 0 & 1 & -1 & 0 & 1 \\ 0 & 1 & 0 & 0 & 0 & 1 & -1 & 0 \\ 0 & 0 & 1 & 0 & 0 & 0 & 1 & -1 \\ 0 & 0 & 0 & 1 & -1 & 0 & 0 & 1 \end{pmatrix}$$

From the set of equations (1),

$$\begin{pmatrix} v_1 \\ v_2 \\ v_3 \\ v_4 \\ v_5 \\ v_6 \\ v_7 \\ v_8 \end{pmatrix} \begin{matrix} = \\ = \\ = \\ = \\ = \\ = \\ = \\ = \end{matrix} \begin{pmatrix} 1 & 0 & 0 & 0 \\ 0 & 1 & 0 & 0 \\ 0 & 0 & 1 & 0 \\ 0 & 0 & 0 & 1 \\ 1 & 0 & 0 & -1 \\ -1 & 1 & 0 & 0 \\ 0 & -1 & 1 & 0 \\ 0 & 0 & -1 & 1 \end{pmatrix} \begin{pmatrix} e_1 \\ e_2 \\ e_3 \\ e_4 \end{pmatrix}$$

In compact form,

$$[V_b]=[Q]^T[V_T] \qquad \text{...(3)}$$

Where,

$[V_b]$ = Branch Voltage Column Matrix of order b × 1.

$[V_T]$ = Tree Branch Voltage Column Matrix of order n × 1.

$[Q]$ = Cut-Set Matrix of order n × b.

$$\begin{pmatrix} 1 & 0 & 0 & 0 & 1 & -1 & 0 & 0 \\ 0 & 1 & 0 & 0 & 0 & 1 & -1 & 0 \\ 0 & 0 & 1 & 0 & 0 & 0 & 1 & -1 \\ 0 & 0 & 0 & 1 & -1 & 0 & 0 & 1 \end{pmatrix} \begin{pmatrix} j_1 \\ j_2 \\ j_3 \\ j_4 \\ j_5 \\ j_6 \\ j_7 \\ j_8 \end{pmatrix} = 0$$

In compact form,

$$[Q]\,[I_b]=0 \qquad \text{...(4)}$$

Where,

$[I_b]$ = Branch Current Column Matrix of order b × 1.

$[Q]$ = Cut-Set Matrix of order n × b.

Twig Voltages

Relation between Incidence Matrix and f-cut-set Matrix

We know that KCL equations can be obtained from a graph by using incidence as well as f-cut-set matrices and are given as,

$$A\,I_b = 0 \quad \text{...(5)}$$

And,

$$Q_f\,I_b = 0 \quad \text{...(6)}$$

Let,

$$I_b = \begin{bmatrix} I_t \\ I_l \end{bmatrix} \quad \text{...(7)}$$

Therefore, the above equation are written as,

$$[A_t, A_l]\begin{bmatrix} I_t \\ I_l \end{bmatrix} = 0$$

And,

$$[Q_t, Q_l]\begin{bmatrix} I_t \\ I_l \end{bmatrix} = 0$$

or,

$$[I, Q_l]\begin{bmatrix} I_t \\ I_l \end{bmatrix} = 0$$

$$A_t I_t + A_l I_l = 0$$

or,

$$I_t = -A_t^{-1} A_l I_l \quad \text{...(8)}$$

Whereas from the f-cut-matrix we have,

$$I_t = -Q_l I_l \quad ...(9)$$

Hence,

$$Q_l = A_t^{-1} A_l \quad ...(10)$$

This means that the rows of f-cut-set matrix are a linear combination of incidence matrix. Therefore, if incidence matrix for a tree is given, the f-cut-set matrix can be obtained.

Relation between f-loop and f-cut-set Matrices

The relation between f-loop matrix and f-cut-set matrix hence can be obtained from the two equations $B^T{}_t = -A_t^{-1} A_l$ and $Q_l = A_t^{-1} A_l$,

$$Q_l = -B^T{}_t \quad ...(11)$$

Now since,

$$B_f = [B_t I] = \left[-Q^T{}_l, I\right] \quad ...(12)$$

And,

$$Q_f = [Q_t, Q_l] = [I, Q_l] \quad ...(13)$$

Hence,

$$B_f Q_f^T = \left[-Q_l^T, I\right]\begin{bmatrix} I \\ Q_l^T \end{bmatrix} = -Q_l^T + Q_l^T = 0 \quad ...(14)$$

Similarly,

$$Q_f B_f^T = [I, Q_l]\begin{bmatrix} -Q_l \\ I \end{bmatrix} = -Q_l + Q_l = 0 \quad ...(15)$$

This shows that f-cut-set and f-loop matrices are also orthogonal to each other.

It can be seen from previous manipulations that when 2, 4, 5 are taken as twigs $Q_l = -B_t^T = A_t^{-1} A_l$. Similarly for any graph for a particular tree these relations hold good which means if any one of the three matrices is known, the other two can be evaluated using these relations. If the system to be analyzed is large the formulation of these matrices is very simple and if one of the matrices is known other large tree matrices can be handled with ease on digital computers as standard program on matrix manipulation are readily available.

As it is evident from equations, (10) and (11) if we are interested in all the branch currents the minimum number of independent currents correspond to the link currents. Once link currents are known twig currents can be obtained through the use of either equation (10) (i.e., incidence matrix) or equation (11) i.e., f-cut set matrix. Hence the currents through all the branches can be obtained. We have also seen that in order to obtain branch voltages in a network, the minimum number of independent variables is the twig voltages. Once twig voltages are known, then through the use of the following relation all branch voltages can be obtained,

$$V_b = Q_f^T V_t \qquad \text{...(16)}$$

In general for a network with n nodes and b branches KVL equations are written as,

$$B_f V_b = 0 \qquad \text{...(17)}$$

Where Bf is the fundamental loop matrix of order (b - n + 1) x $_b$ and V_b is the column matrix of b x 1. Now partitioning B_f into B_l and B_t, we have,

$$[B_t, I]\begin{bmatrix} V_t \\ V_l \end{bmatrix} = 0 \qquad \text{...(18)}$$

or,

$$B_t V_t + V_l = 0$$

or,

$$V_l = -B_t V_t \qquad \text{...(19)}$$

Therefore, if V_t is known i.e., if the twig voltage are known, the remaining branches of the network which are links, their voltages can be obtained using the above relation. Therefore, to obtain branch voltages of the network, the minimum number of independent variables equals the twig voltages.

We know that from equation (13),

$$Q_l = -B_t^T$$

or,

$$B_t = -Q_l^T$$

$$V_l = +Q_l^T V_t$$

Now,

$$V_b = \begin{bmatrix} V_t \\ V_l \end{bmatrix} = \begin{bmatrix} U & V_t \\ Q_l^T & V_t \end{bmatrix} = \begin{bmatrix} U \\ Q_l^T \end{bmatrix} V_t$$

$$= Q_f^T V_t \quad \text{...(20)}$$

This means the branch voltages V_b are a linear combination of twig voltages. Thus there are independent variables corresponding to twig voltages.

We can also obtain the branch voltages using nodal voltages as independent variables through the use of incidence matrix and is given as;

$$V_b = A^T V_n \quad \text{...(21)}$$

Where V_b the branch voltage and AT is the transpose of the incidence matrix and V_n is the voltages of the various nodes.

We know that the twig currents are a linear combination of link currents through the use of f-cut-set matrix equation (11) and is given as;

$$I_t = -Q_l I_l = B_t^T I_l$$

Also,

$$I_l = U I_l$$

Therefore,

$$I_b = \begin{bmatrix} I_t \\ I_l \end{bmatrix} = \begin{bmatrix} B_t^T & I_l \\ U & I_l \end{bmatrix} = \begin{bmatrix} B_t^T \\ U \end{bmatrix} I_l$$

$$= B_f^T I_l \quad \text{...(22)}$$

This shows that the branch currents are expressed as a linear combination of link currents.

Similarly branch currents can be expressed as a linear combination of mesh currents and is given as,

$$I_b = B_m^T I_m \quad \text{...(23)}$$

Where I_m is the mesh current column vector.

The observations made above are summarized for ready reference.

If b is the number of branches and n the number of nodes then the dimensions of various matrices are,

$$A \rightarrow (n-1) \times b$$

$$B_f \rightarrow (b-n+1) \times b$$

$$Q_f \rightarrow (n-1) \times b$$

The number of trees in a graph are given by,

$$\det \left[A \ A^T \right]$$

When these matrices are partitioned corresponding to twigs and links.

- A_t and A_l are non-unit matrices.
- B_t is non-unit. B_l is unit matrix.
- Q_t is unit matrix. Q_l is non-unit matrix.

Twig currents and link currents are related as,

$$I_t = -A_t^{-1} A_l I_l = -Q_l I_l = B_t^T I_l$$

The branch currents in terms of link currents are given as,

$$I_b = B_f^T I_l$$

The link voltages ae specified as a function of twig voltages as,

$$V_l = -B_t V_t = Q_l^T V_t$$

And the branch voltages are given as,

$$V_b = Q_f^T V_t$$

Branch voltages in terms of node voltages are given as,

$$V_b = A^T V_n.$$

2

Steady State Analysis of AC Circuits

2.1 AC Circuits and its Response to Sinusoidal Excitation

Many important applications in electrical engineering involves AC (sinusoidal) currents and voltages. Electric power is distributed worldwide using AC signals with a frequency of either 50 or 60 Hz (60 Hz in the United States).

The transmission and reception of electromagnetic waves necessary in any wireless application involves AC signals. Even non-AC signals (such as digital signals) are commonly analyzed using Fourier analysis which is based on the fact that any time-variant signal can be expressed as the superposition of AC signals.

When AC signals are initially applied to electric circuits (usually via a switch), transient responses are produced in the circuit that decay to zero over time. Circuit responses determined using steady-state AC analysis are those found after the transients responses have decayed to zero.

AC Circuit Analysis

The mathematics associated with AC circuit analysis can be greatly simplified by transforming the instantaneous sinusoidal signals (in the time-domain) into complex-valued phasors (in the frequency domain). The real-valued instantaneous voltage and current [v(t) and i(t)] are transformed into the complex-valued phasor voltage and current [V and I] where we use boldface letters to denote a complex-valued phasor.

Time – domain (instantaneous quantities)		Frequency - domain Phasors
$v(t)=V_{peak}\cos(\omega t+\theta_v)$	$\Leftrightarrow$	$\mathbf{V}=V_{peak}\angle\theta_v$
$i(t)=I_{peak}\cos(\omega t+\theta_i)$		$\mathbf{I}=I_{peak}\angle\theta_i$

We will perform the AC circuit analysis in the frequency-domain and transform the results back into the time-domain. The primary advantage to performing the circuit analysis in the frequency-domain is the elimination of derivatives and integrals of currents and voltages which occur in the time-domain for inductors and capacitors. These time-domain derivatives and integrals transform into simple algebraic expressions in the frequency-domain.

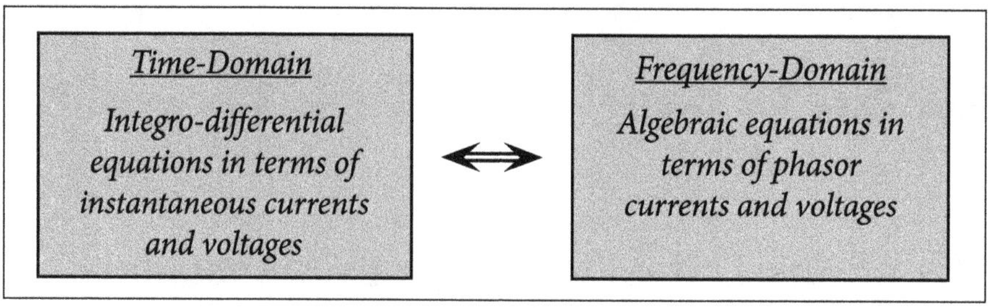
Time-Domain
Integro-differential equations in terms of instantaneous currents and voltages
Frequency-Domain
Algebraic equations in terms of phasor currents and voltages

The transformation of a time-domain signal to a frequency-domain signal is based on Euler's identity,

$$e^{j\omega t} = \cos(\omega t) + j\sin(\omega t) \qquad \text{(Euler's identity)}$$

A phasor voltage or current is completely described by its magnitude (V_{peak} or I_{peak}) and phase (θ_v or θ_i) in the complex plane. Alternatively, phasors can be written in terms of real and imaginary components.

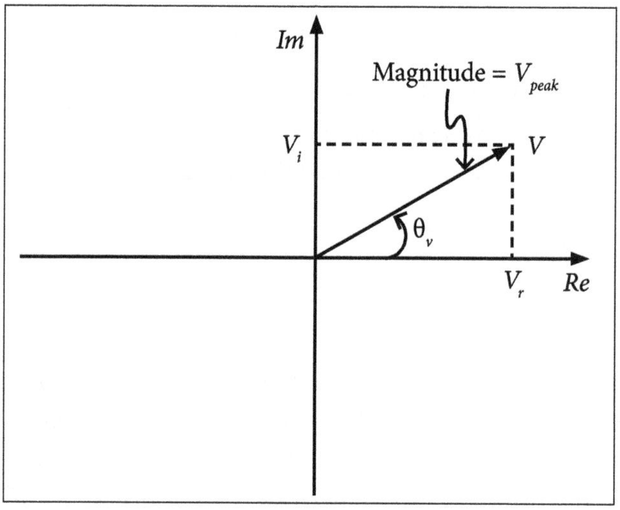

Characteristics graph.

The phasor voltage V in the figure above can be written in any of the following forms:

$$V = V_{peak}\angle\theta_v \qquad \text{(magnitude and phase)}$$

$$= V_r + jV_i \qquad \text{(real + imaginary)}$$

$$= V_{peak}\cos\theta_v + jV_{peak}\sin\theta_v$$

$$= V_{peak}\left[\cos\theta_v + j\sin\theta_v\right]$$

$$=V_{peak}\, e^{j\theta_v}$$

When manipulating phasors, certain mathematical operations are performed more efficiently in magnitude/phase form (multiplication and division of phasors) while others are performed more efficiently in real + imaginary form (addition and subtraction of phasors).

Assume that we are given two arbitrary phasors V_1 and V_2 as shown in the figure below,

$$V_1 = V_1 \angle\theta_1 = V_{1r} + jV_{1i}$$

$$V_2 = V_2 \angle\theta_2 = V_{2r} + jV_{2i}$$

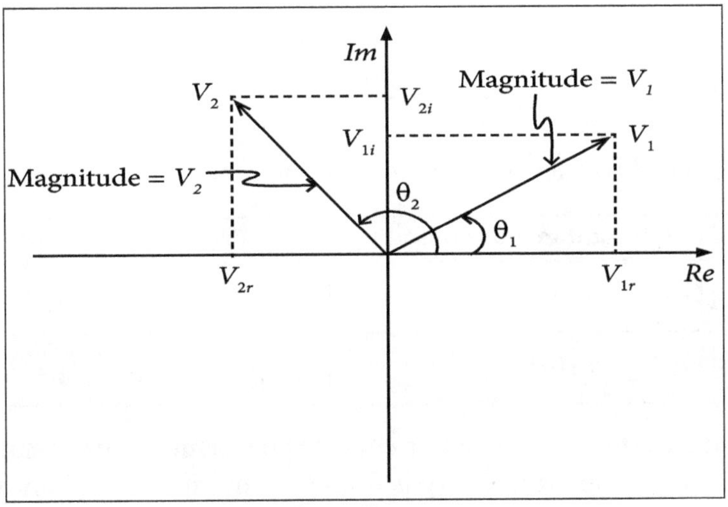

Phasor diagram.

Addition (use real + imaginary),

$$V_1 + V_2 = (V_{1r} + jV_{1i}) + (V_{2r} + jV_{2i}) = (V_{1r} + V_{2r}) + j\,(V_{1i} + V_{2i})$$

Subtraction (use real + imaginary),

$$V_1 - V_2 = (V_{1r} + jV_{1i}) - (V_{2r} + jV_{2i}) = (V_{1r} - V_{2r}) + j\,(V_{1i} - V_{2i})$$

Multiplications (use magnitude/phase),

$$V_1 V_2 = (V_1 \angle\theta_1)(V_2 \angle\theta_2) = V_1 V_2 \angle(\theta_1 + \theta_2)$$

Division (uses magnitude/phase),

$$\frac{V_1}{V_2} = \frac{V_1 \angle\theta_1}{V_2 \angle\theta_2} = \frac{V_1}{V_2} \angle(\theta_1 - \theta_2)$$

Given:

The general expressions for the instantaneous AC voltage is,

$$V_1 = V_1 \angle\theta_1 = V_{1r} + jV_{1i}$$

$$V_2 = V_2 \angle\theta_2 = V_{2r} + jV_{2i}$$

The transformation of the instantaneous quantities into phasors by applying Euler's identity, current v(t), i(t) is defined by,

$$v(t) = V_{peak}\cos(\omega t + \theta_v)$$
$$= Re\left[V_{peak}\,e^{j(\omega t+\theta_v)}\right] = Re\left[\left(V_{peak}\,e^{j\theta_v}\right)e^{j\omega t}\right] = Re\left[Ve^{j\omega t}\right]$$

$$i(t) = I_{peak}\cos(\omega t + \theta_i)$$
$$= Re\left[I_{peak}\,e^{j(\omega t+\theta_i)}\right] = Re\left[\left(I_{peak}\,e^{j\theta_i}\right)e^{j\omega t}\right] = Re\left[Ie^{j\omega t}\right]$$

The time-domain and frequency-domain signals are related by,

Time – domain (instantaneous quantities)		**Frequency - domain (Phasors)**
$v(t) = V_{peak}\cos(\omega t + \theta_v)$	$\Leftrightarrow$	$V = V_{peak}\angle\theta_v$
$i(t) = I_{peak}\cos(\omega t + \theta_i)$		$I = I_{peak}\angle\theta_i$

Such that given an arbitrary voltage or current phasor in the frequency-domain, the corresponding instantaneous quantity is found by multiplying the phasor by em and taking the real part of the product.

$$v(t) = Re\left[Ve^{j\omega t}\right]$$

$$i(t) = Re\left[Ie^{j\omega t}\right]$$

2.1.1 Response to Sinusoidal Excitation: Pure Resistance, Pure Inductance and Pure Capacitance

Pure Resistance

In case of resistor, the voltage and the current are in same phase or we can say that the phase angle difference between voltage and current is zero.

According to Ohm's Law, the instantaneous voltage V across a resistor is proportional to the instantaneous current I flowing through it.

$$V = R.I$$

$$V = R.\ I_m \sin\omega t$$

$$V = V_m \sin\omega t$$

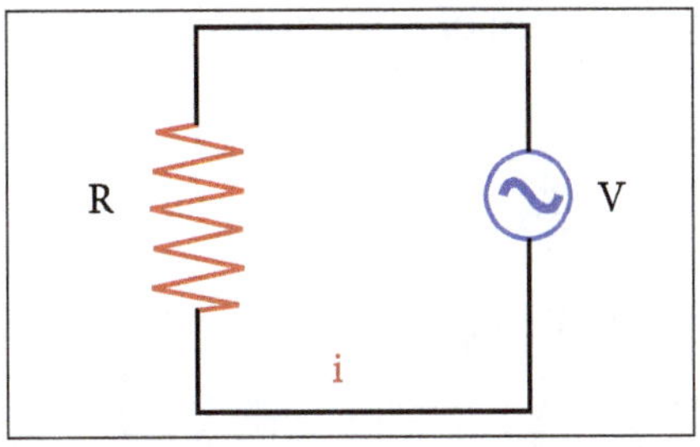

Resistor in an AC circuit.

The two graphs show the voltage across the resistor and the current flowing through the resistor as a function of time.

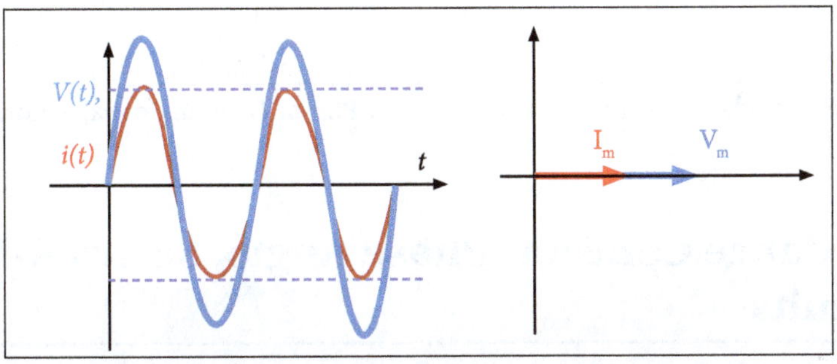

Sinusoidal variation in phase V and I & Phasor Diagram for resistor.

Thus voltage and current are in phase.

Pure Inductor

In inductor, the voltage and the current are not in phase. The voltage leads that of current by 90° or in the other words, voltage attains its maximum and zero value 90° before the current attains it.

$$V_m = X_L . I_m$$

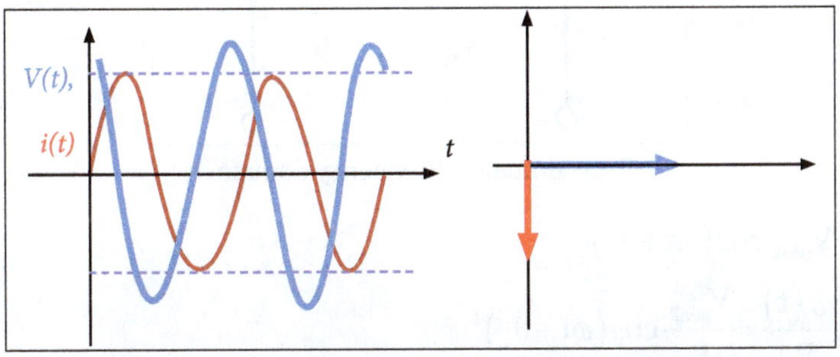

Sinusoidal variation in phase V and I & Phasor Diagram for inductor.

Pure Capacitor

In case of capacitor, the current leads the voltage by 90° or in the other words, voltage attains its maximum and zero value 0° after the current attains it, i.e., the phasor diagram of capacitor is exactly opposite of inductor.

$$V_m = X_C \cdot I_m$$

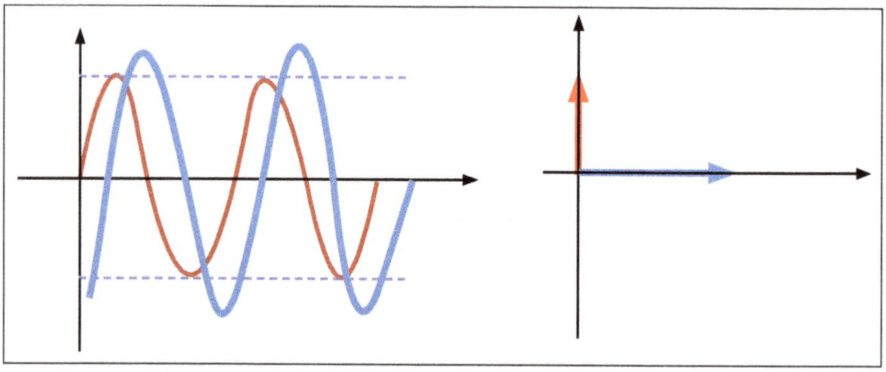

Sinusoidal variation in phase V and I & Phasor Diagram for capacitor.

2.2 Impedance Concept, Phase Angle, Series R-L, R-C and R-L-C Circuits

The impedance of a circuit element is defined as the ratio of the phasor voltage across the element to the phasor current through the element (units = o). We can determine the impedance of resistors, capacitors and inductors using the time-domain terminal characteristics for each element. Resistance impedance:

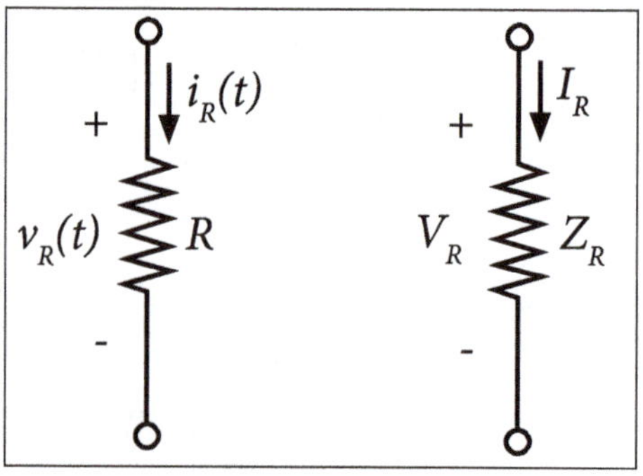

Time domain - Frequency domain.

$$v_R(t) = V_{peak}\cos(\omega t + \theta_v)$$

$$i_R(t) = \frac{v_R(t)}{R} = \frac{V_{peak}}{R}\cos(\omega t + \theta_v)$$

$$V_R = V_{peak} \angle \theta_v$$

$$I_R = \frac{V_{peak}}{R} \angle \theta_v$$

$$Z_R = \frac{V_R}{I_R} = \frac{V_{peak} \angle \theta_v}{\frac{V_{peak}}{R} \angle \theta_v}$$

$$Z_R = R$$

Such that the resistor voltage and current are in phase (same phase angle). The resistor impedance is purely real and is independent of frequency.

Capacitor Impedance

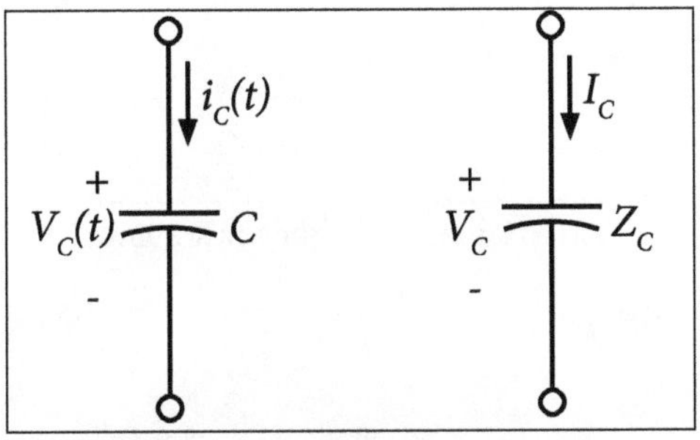

Time domain - Frequency domain.

$$v_C(t) = V_{peak} \cos(\omega t + \theta_v)$$

$$i_C(t) = C \frac{dv_C(t)}{dt}$$

$$= C\left[-\omega V_{peak} \sin(\omega t + \theta_v)\right]$$

$$= \omega C V_{peak} \cos(\omega t + \theta_v + 90°)$$

$$V_C = V_{peak} \angle \theta_v$$

$$I_C = \omega C V_{peak} \angle(\theta_v + 90°)$$

$$Z_C = \frac{V_C}{I_C}$$

$$= \frac{V_{peak} \angle \theta_v}{\omega C V_{peak} \angle(\theta_v + 90°)}$$

$$= \frac{1 \angle -90°}{\omega C} = \frac{1}{j \omega C}$$

Such that the capacitor voltage and current are out of phase by 90° (the capacitor current always leads the capacitor voltage by 90°). The capacitor impedance is purely imaginary and is inversely proportional to frequency. The capacitor impedance is very large at low frequencies and very small at high frequencies.

Inductor Impedance

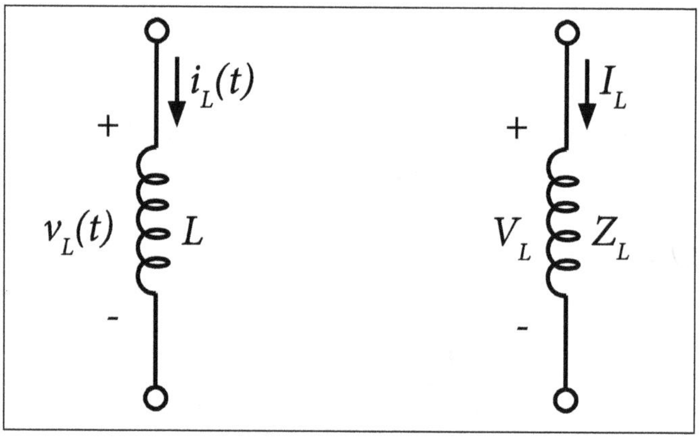

Time domain - Frequency domain.

$$i_L(t) = I_{peak}\cos(\omega t + \theta_i)$$

$$v_L(t) = L\frac{di_L(t)}{dt}$$

$$= L\left[-\omega I_{peak}\sin(\omega t + \theta_i)\right]$$

$$= \omega L I_{peak}\cos(\omega t + \theta_v + 90°)$$

$$I_L = I_{peak}\angle\theta_i$$

$$V_L = \omega L I_{peak}\angle(\theta_v + 90°)$$

$$Z_L = \frac{V_L}{I_L}$$

$$= \frac{\omega L I_{peak}\angle(\theta_v + 90°)}{I_{peak}\angle\theta_v}$$

$$= \omega L\angle 90° = j\omega L$$

Such that the inductor voltage and current are out of phase by 90° (the inductor current always lags the inductor voltage by 90°). The inductor impedance purely imaginary and is directly proportional to frequency. The inductor impedance is very small at low frequencies and very large at high frequencies.

Impedance and Phase Angle

The phase angle is the phase difference between the total current and the source voltage.

The impedance of a series RC circuit is determined by both the resistance (R) and the capacitive reactance (X_C).

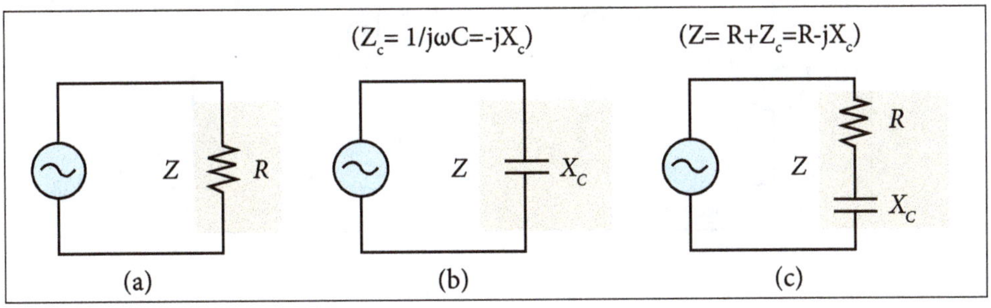

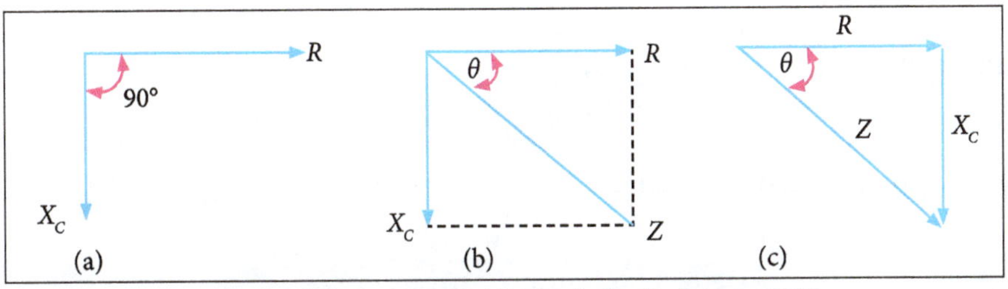

(a) Z =R, (b) Z = XC and (c) Z includes both R and XC.

In the series RC circuit, the total impedance is the phasor sum of R and jX_C.

Impedance magnitude, $Z = \sqrt{R^2 + X^2C}$

Phase angle, $\theta = \tan^{-1}(X_C / R)$

Problems

1. Let us determine the impedance and the phase angle of series RC circuit for the following figure.

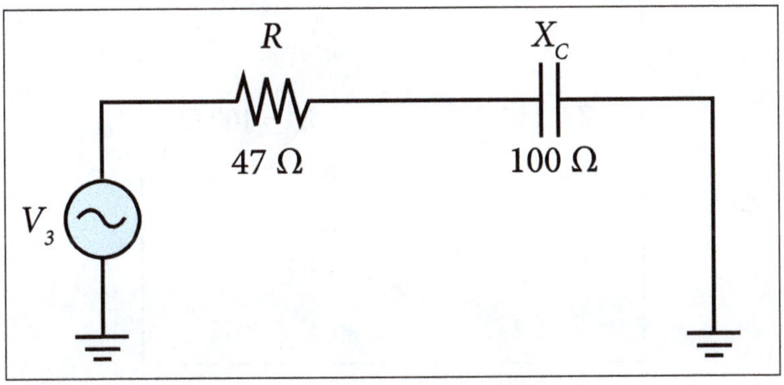

Solution:

Given:

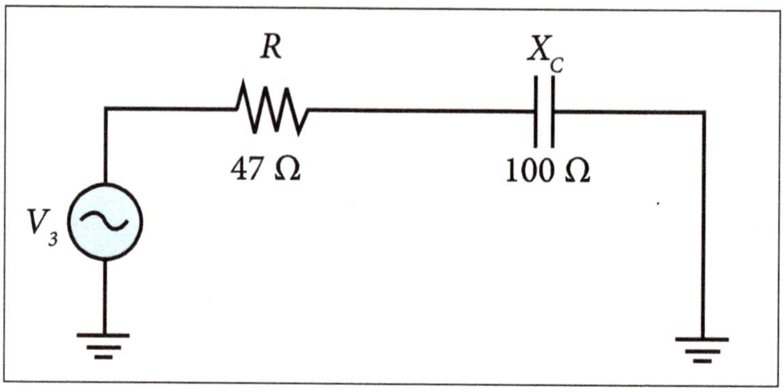

Formula:

$$\text{Impedance, } |Z| = \sqrt{R^2 + (X_L - X_C)^2}$$

$$\text{Phase angle, } \tan\theta = \frac{X_L - X_C}{R}$$

Here,

$$X_L = 0$$

$$Z = \sqrt{(47)^2 + (100)^2} = 110\Omega$$

$$\theta = \tan^{-1}(100/47) = \tan^{-1}(2.13) = 64.8$$

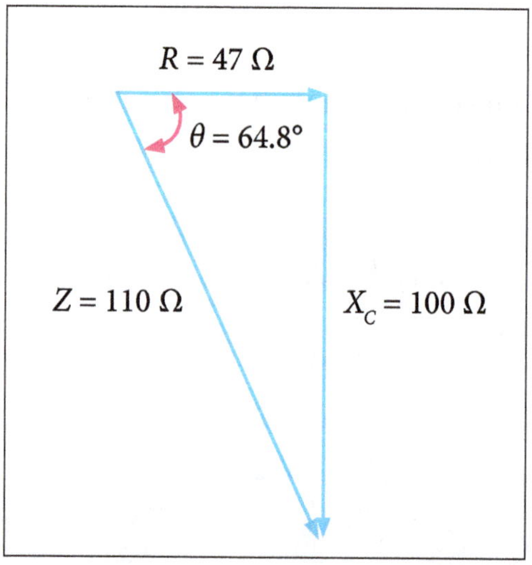

2. If the current is 0.2 mA, let us determine the source voltage and the phase angle for the series RC circuit shown below.

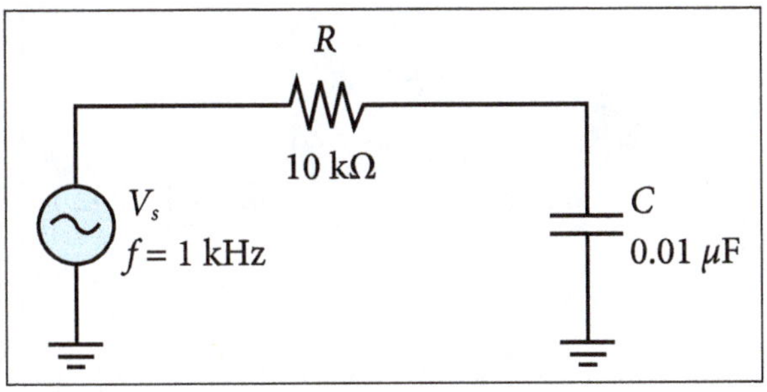

Solution:

Given:

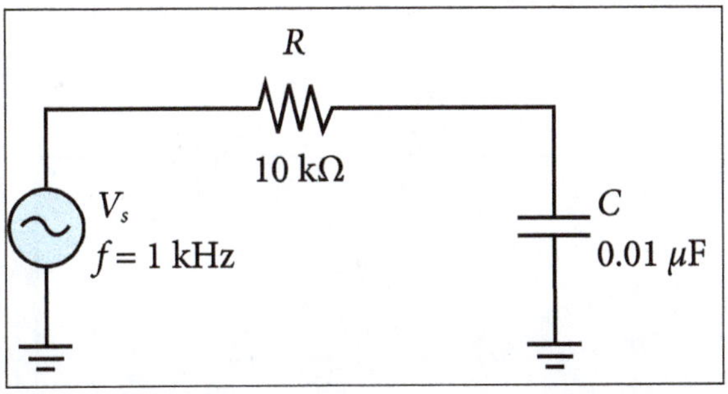

Formula:

Reactive capacitance, $X_C = 1/2\pi RC$

Impedance, $|Z| = \sqrt{R^2 + (X_L - X_C)^2}$

Phase angle, $\tan\theta = \dfrac{X_L - X_C}{R}$

Here,

$X_L = 0$

$X_C = 1/2\pi\left(1\times10^3\right)\left(0.01\times10^{-6}\right) = 15.9\,k\Omega$

$Z = \sqrt{\left(10\times10^3\right)^2 + \left(15.9\times10^3\right)^2} = 18.8\,k\Omega$

$V_S = IZ = (0.2\,mA)(18.8\,k\Omega) = 3.76\,V$

$\theta = \tan^{-1}(15.9k/10k) = 57.8$

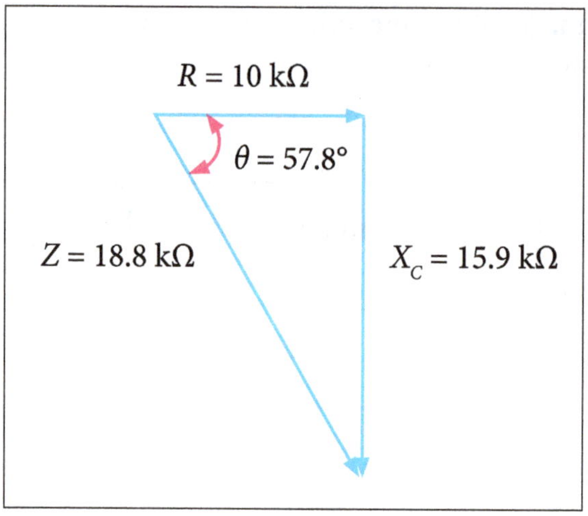

3. Let us determine the impedance and the phase angle for the RL circuit shown below.

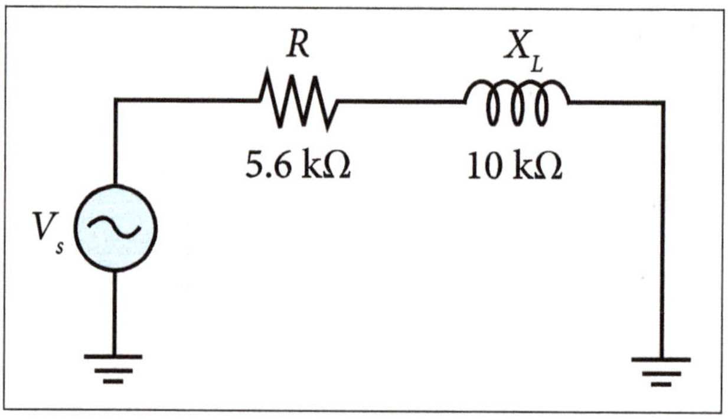

Solution:

Given:

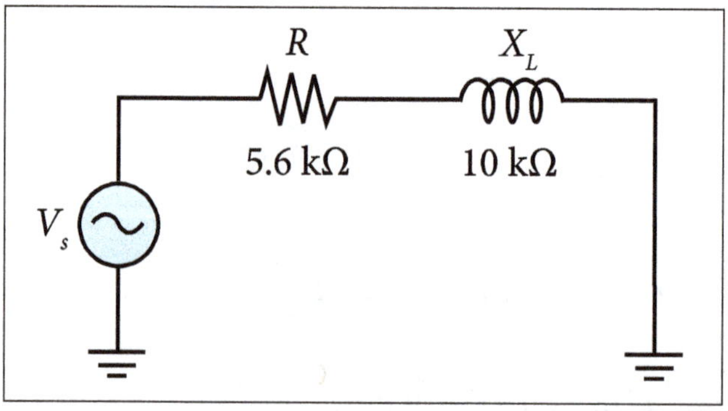

$$Z = \sqrt{(5.6)^2 + (10)^2} = 11.5\,\text{k}\Omega$$

$$\theta = \tan^{-1}(10/5.6) = \tan^{-1}(1.786) = 60.8°$$

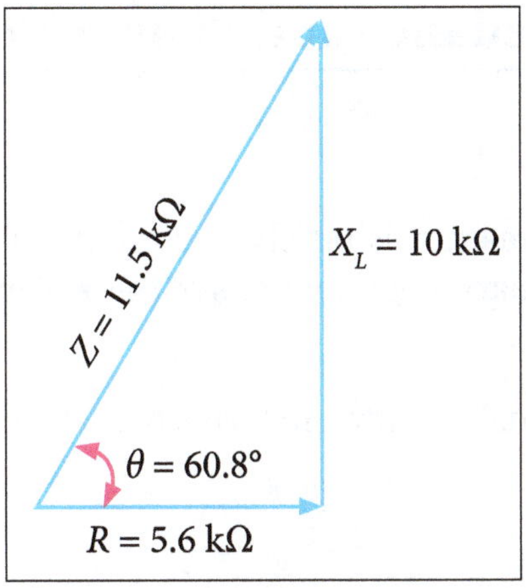

4. If the current is 0.2 mA, let us determine the source voltage and the phase angle for the RL circuit shown below.

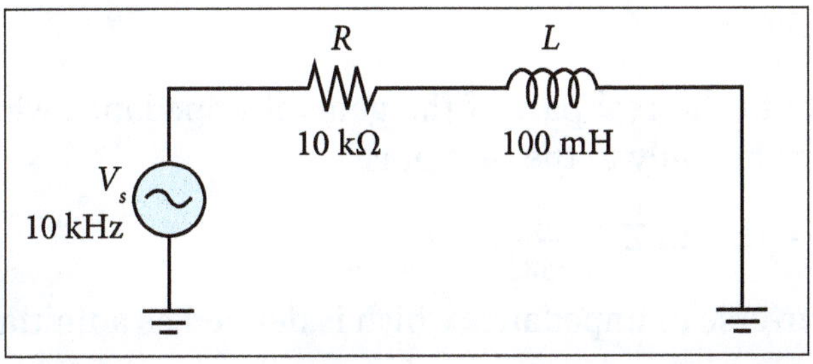

Solution:

Given:

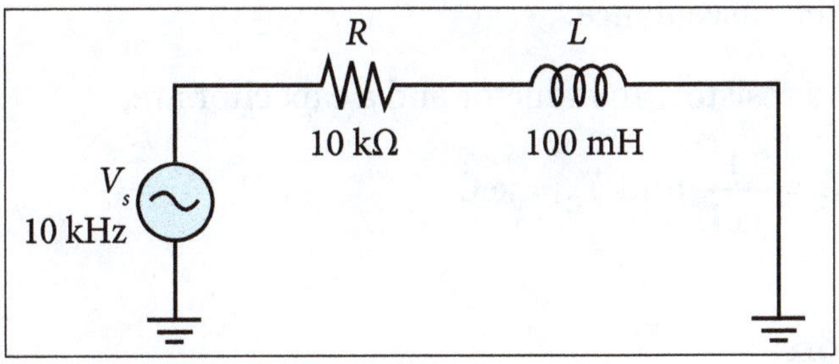

$$X_L = 2\pi\left(10\times10^3\right)\left(100\times10^{-3}\right)=6.28/k\Omega$$

$$Z=\sqrt{\left(10\times10^3\right)^2+\left(6.28\times10^3\right)^2}=11.8k\Omega$$

$$V_S=IZ=(200\mu A)(11.8k\Omega)=2.36V$$

$$\theta=\tan^{-1}(6.28k/10k)=34.2°$$

2.3 Complex Impedance and Phasor Notation

Complex Impedance

While the impedance of a resistor is purely real and the impedance of inductors and capacitors are purely imaginary, combinations of these elements produces impedances that are, in general, complex.

The general impedance Z can be written in terms of real and imaginary components as,

$$Z=\frac{V}{I}=R+jX(\Omega)$$

Where,

R is the resistance.

X is the reactance.

Resistors contribute to the real part of the general impedance while ideal capacitors and inductors contribute only to the reactance.

$$Z_R=R,\ Z_L=j\omega L \text{ and } Z_C=\frac{1}{j\omega C}$$

We can define the inverse of impedance which is defined as admittance (Y).

$$Y=\frac{1}{Z}=\frac{1}{V}=G+jB(\mho \text{ or } S)$$

The real part of the admittance (G) is the conductance while the imaginary part of the admittance (B) is the susceptance.

The admittance of a resistor, an inductor and a capacitor are,

$$Y_R=\frac{1}{R},\ Y_L=\frac{1}{j\omega L} \text{ and } Y_C=j\omega C$$

Phasor Notation

It is the process of constructing a single complex number that has the amplitude and phase angle of the given sinusoidal waveform. Then the phasor notation or phasor transform transfers the sinusoidal function of $A(t)=A_m\sin(\omega t\pm\phi)$ from the time domain into the complex number domain which is also termed as frequency domain.

For example:

$$V_{(t)}=V_m\sin(\omega t+\theta) \quad \Leftrightarrow \quad \text{Euler's identity}: e^{\pm j\theta}=\cos\theta\pm j\sin\theta$$

$$V_{(t)} = 20\sin(\omega t + 30^\circ)\,\text{Volts} \Rightarrow \left(\frac{20}{\sqrt{2}}\right)\angle 30^\circ \Rightarrow V_{RMS} = 14.14\angle 30^\circ$$

or,

$$V_{(t)} = 35\sin(\omega t + 45^\circ)\,\text{Volts} \Rightarrow \left(\frac{35}{\sqrt{2}}\right)\angle 45^\circ \Rightarrow V_{RMS} = 24.75\angle 45^\circ$$

or,

$$V_{(t)} = 10\sin(\omega t - 30^\circ)\,\text{Volts} \Rightarrow \left(\frac{10}{\sqrt{2}}\right)\angle -35^\circ \Rightarrow V_{RMS} = 7.07\angle -30^\circ$$

or,

$$V_{(t)} = 20\sin(\omega t)\,\text{Volts} \Rightarrow \left(\frac{20}{\sqrt{2}}\right)\angle 0^\circ \Rightarrow V_{RMS} = 14.14\angle 0^\circ$$

Note that √2 converts the maximum amplitude into an effective RMS value with the phase angle given in radians.

Problems

1. Let us determine i_x in the circuit of figure shown below using nodal analysis.

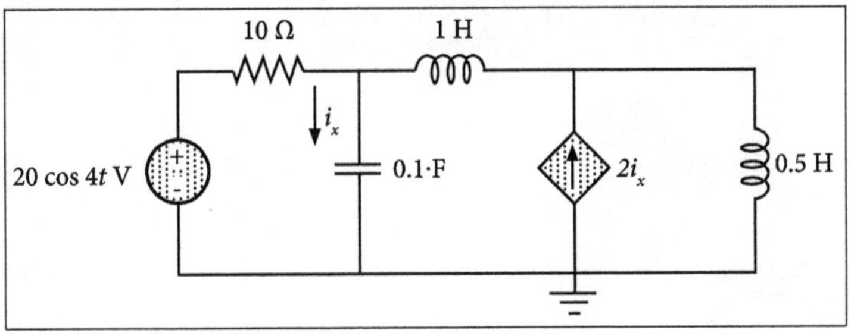

Solution:

Given:

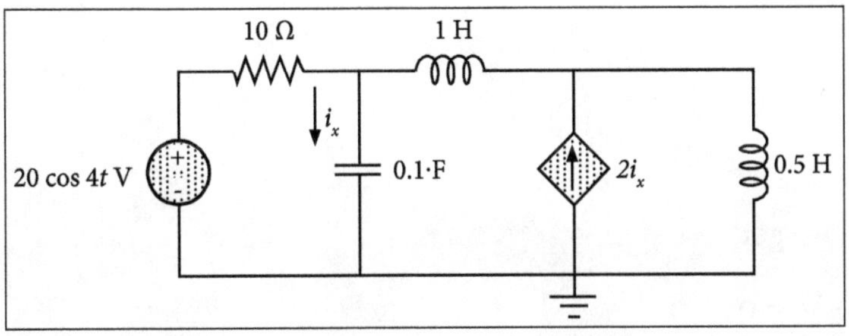

We first convert the circuit to the frequency domain,

$20\cos 4t \Rightarrow 20\angle 0°, \qquad \omega = 4\,rad/s$

$1H \Rightarrow j\omega L = j4$

$0.5 \Rightarrow j\omega L = j2$

$0.1\,F \Rightarrow \frac{1}{j\omega C} = -j2.5$

Thus, the frequency-domain equivalent circuit is shown below.

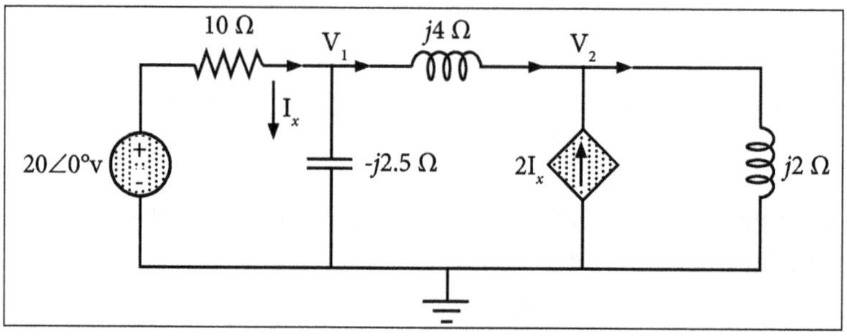

Applying KCL at node 1,

$$\frac{20-V_1}{10} = \frac{V_1}{-j2.5} + \frac{V_1 - V_2}{j4}$$

$$(1+j1.5)V_1 + j2.5V_2 = 20 \qquad \text{...(1)}$$

$$2I_x + \frac{V_1 - V_2}{j4} = \frac{V_2}{j2}$$

But,

$$I_x = V_1 / -j2.5$$

$$\frac{2V_1}{-j2.5} + \frac{V_1 - V_2}{j4} = \frac{V_2}{j2}$$

By simplifying, we get,

$$11V_1 + 15V_2 = 0 \qquad \text{...(2)}$$

Equation (1) and (2) can be put in matrix form as,

$$\begin{bmatrix} 1+j1.5 & j2.5 \\ 11 & 15 \end{bmatrix}\begin{bmatrix} V_1 \\ V_2 \end{bmatrix}=\begin{bmatrix} 20 \\ 0 \end{bmatrix}$$

We obtain the determinants as,

$$\Delta=\begin{vmatrix} 1+j1.5 & j2.5 \\ 11 & 15 \end{vmatrix}=15-j5$$

$$\Delta_1=\begin{vmatrix} 20 & j\,2.5 \\ 0 & 15 \end{vmatrix}=300,\quad \Delta_2=\begin{vmatrix} 1+j1.5 & 20 \\ 11 & 0 \end{vmatrix}=-220$$

$$V_1=\frac{\Delta_1}{\Delta}=\frac{300}{15-j5}=18.97\angle 18.43°\text{V}$$

$$V_2=\frac{\Delta_2}{\Delta}=\frac{-220}{15-j5}=13.91\angle 198.3°\text{V}$$

The current I_x is given by,

$$I_x=\frac{V_1}{-j2.5}=\frac{18.97\angle 18.43°}{2.5\angle -90°}=7.59\angle 108.4°\text{A}$$

Transforming this to time domain

$$i_x=7.59\cos(4t+108.4°)\text{A}$$

2. Using node analysis, let us find the node voltages and the currents through all the resistors for the circuit shown in figure below.

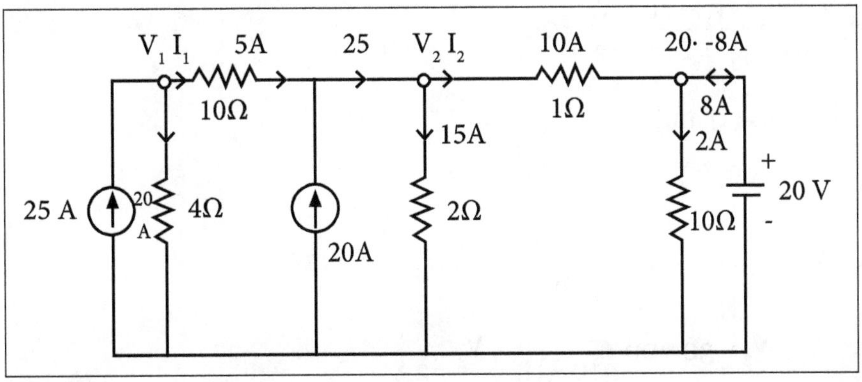

Solution:

Given:

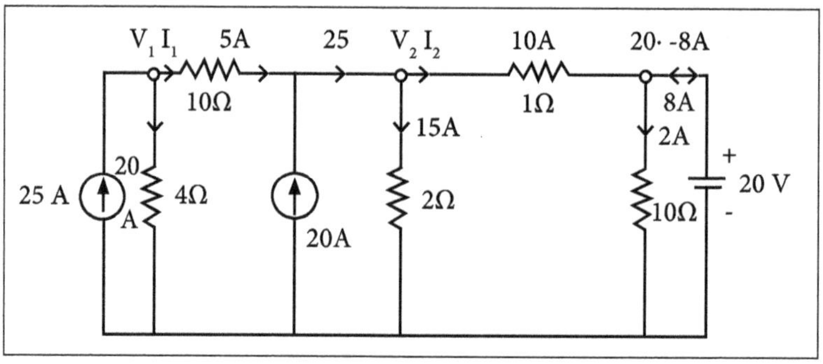

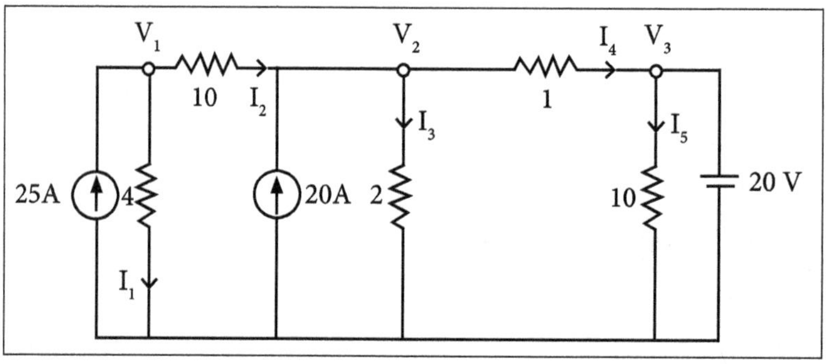

Apply KCL at node "V_1",

$$\frac{V_1 - V_2}{10} + \frac{V_1}{4} = 25 \quad ...(1)$$

Apply KCL at node "V_2",

$$\frac{V_2 - V_1}{10} + \frac{V_2 - V_3}{1} + \frac{V_2}{2} = 20$$

$$\Rightarrow \frac{V_2 - V_1}{10} + \frac{V_2 - 20}{1} + \frac{V_2}{2} = 20 \quad ...(2)$$

Solving equations (1) and (2),

$V_1 = 80$ V; $V_2 = 30$ V

$$I_1 = V_1 / 4 = 20 = A; I_2 = \frac{V_1 - V_2}{10} = 5A; I_3 = \frac{V_2}{2} = 15A$$

$$I_4 = \frac{V_2 - V_3}{1} = \frac{30 - 20}{1} = 10A; I_5 = \frac{V_3}{10} = 2A$$

3. Let us determine I_0 in the circuit of figure shown below using mesh analysis.

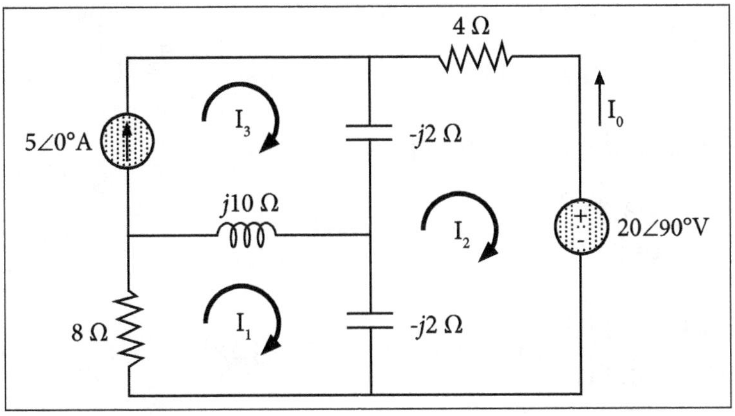

Solution:

Given:

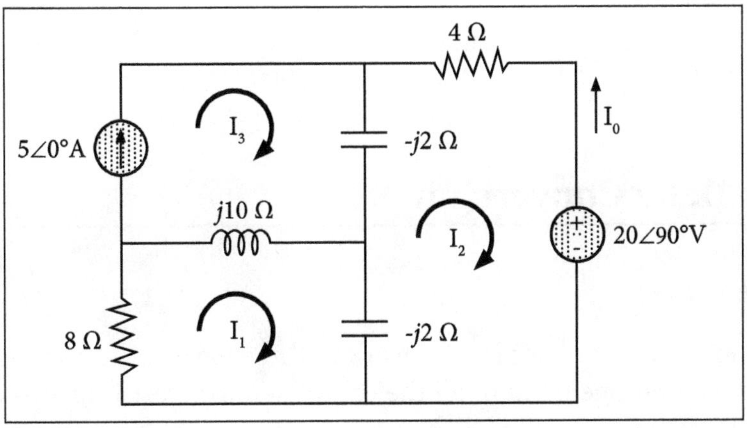

Applying KVL to mesh 1, we obtain as,

$$(8+j10-j2)I_1-(-j2)I_2-j10I_3=0 \quad ...(1)$$

For mesh 2,

$$(4-j2-j2)I_2-(-j2)I_1-(-j2)I_3+20\angle 90°=0 \quad ...(2)$$

For mesh 3, $I_3 = 5$.

Substituting this in equation (1) and (2), we get,

$$(8+j8)I_1+j2I_2=j50 \quad ...(3)$$

$$j2I_1+(4-j4)I_2=-j20-j10 \quad ...(4)$$

Equation (3) and (4) can be put in matrix form as,

$$\begin{bmatrix} 8+j8 & j2 \\ j2 & 4-j4 \end{bmatrix}\begin{bmatrix} I_1 \\ I_2 \end{bmatrix}=\begin{bmatrix} j50 \\ -j30 \end{bmatrix}$$

From which we obtain the determinants as,

$$\Delta=\begin{vmatrix} 8+j8 & j2 \\ j2 & 4-j4 \end{vmatrix}=32(1+j)(1-j)+4=68$$

$$\Delta_2=\begin{vmatrix} 8+j8 & j50 \\ j2 & -j30 \end{vmatrix}=340-j240=416.17\angle-35.22°$$

$$I_2=\frac{\Delta_2}{\Delta}=\frac{416.17\angle-35.22°}{68}=6.12\angle-35.22°A$$

The desired current is,

$$I_o=-I_2=6.12\angle144.78°$$

2.4 Star-Delta Conversion

Star Connection

Star connection or star network is a network of three electrical components in such a composition such that one point of all the components or resistors are connected at a common point.

The star connection of three resistors or branches also looks like the English alphabet "Y". Hence, the Star network or connection is also referred to as "Y" or "Wye" connection.

A star or Y connection can also be rearranged to make it look like the English letter "T" and many electrical circuits use the star connection in it's "T" form.

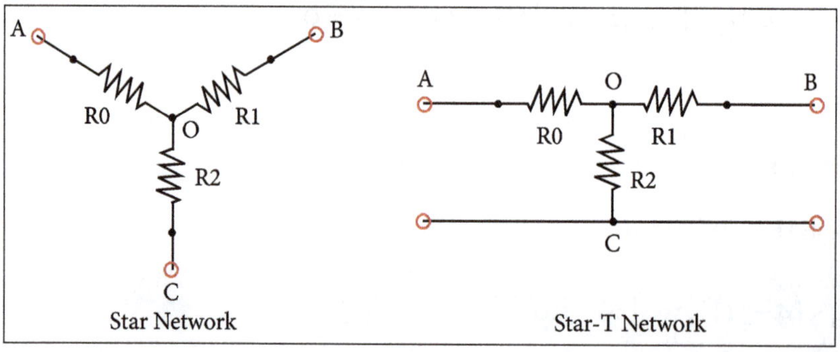

Delta Connection

Delta connection is an electrical network composed of three branches or circuit components, where each branch or element is connected head to tail of other branch or circuit element. The Delta connection is so called because it looks like the Greek alphabet Delta (Δ).

The delta connection can also be re arranged to look like a Pi (Π) symbol so the network is also sometimes referred to as Pi Network.

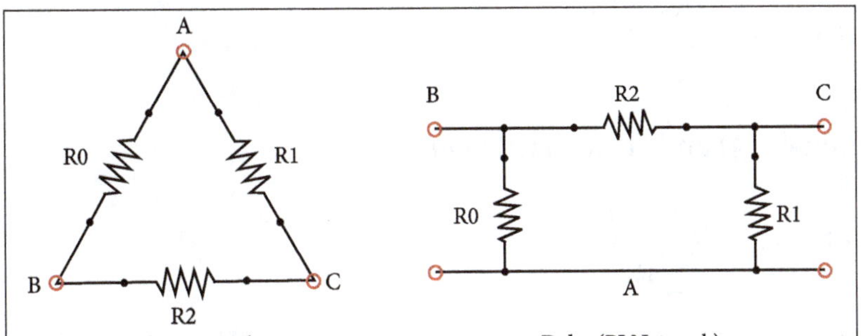

Delta Network   Delta (PI Network)

It is often useful to transform delta network to star and star network to delta while solving an electrical network. The transformation can make the calculations simpler and easier.

Derivation for Star to Delta Conversion

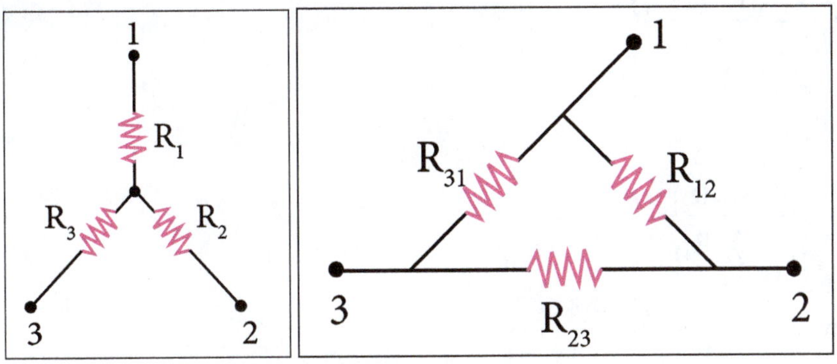

Star-delta conversion

For the two circuits (shown in figure) to be equivalent, the resistance between any two points with the third open should be equal for both the circuits.

∴ Between terminals (1) and (2) with (3) open,

$$R_1 + R_2 = R_{12} + \frac{R_{23}R_{31}}{R_{23} + R_{31}}$$

$$R_1 + R_2 = \frac{R_{12}(R_{23} + R_{31})}{R_{12} + R_{23} + R_{31}}$$

$$R_1 + R_2 = \frac{R_{12}(R_{23} + R_{31})}{\sum R_{12}} \quad ...(1)$$

Between terminals (2) and (3) with (1) open,

$$R_2 + R_3 = \frac{R_{23}(R_{31} + R_{12})}{\sum R_{12}} \quad ...(2)$$

Between terminals (3) and (1) with (2) open,

$$R_3 + R_1 = \frac{R_{31}(R_{12} + R_{23})}{\sum R_{12}} \quad ...(3)$$

By solving (1) – (2) + (3), we get,

$$R_1 - R_2 = \frac{R_{12}(R_{31} - R_{23})}{\sum R_{12}} \quad ...(4)$$

Similarly,

$$2R_1 = \frac{R_{12}(2R_{31})}{\sum R_{12}}$$

Hence,

$$R_1 = \frac{R_{12}R_{31}}{\sum R_{12}}$$

And we have,

$$R_1 = \frac{R_{12}R_{13}}{\sum R_{12}} \quad ...(5)$$

Similarly,

$$R_2 = \frac{R_{21}R_{23}}{\sum R_{12}} \quad ...(6)$$

$$R_3 = \frac{R_{31}R_{32}}{\sum R_{12}} \qquad ...(7)$$

Thus equations (4), (5) and (6) will be used for delta to star conversion.

Now, Multiplying Equation (5) by equation (6), we get,

$$R_1R_2 = \frac{R_{12}^2R_{23}R_{31}}{\left(\sum R_{12}\right)^2} \qquad ...(8)$$

Multiplying Equation (6) by equation (7), we get,

$$R_2R_3 = \frac{R_{12}R_{23}^2R_{31}}{\left(\sum R_{12}\right)^2} \qquad ...(9)$$

Multiplying Equation (7) by equation (5), we get,

$$R_3R_1 = \frac{R_{12}R_{23}R_{31}^2}{\left(\sum R_{12}\right)^2} \qquad ...(10)$$

Substituting Equation (8) and (9) in equation (10), we get,

$$R_1R_2 + R_2R_3 + R_3R_1 = \frac{R_{12}R_{23}R_{31}\left(R_{12} + R_{23} + R_{31}\right)}{\left(\sum R_{12}\right)^2}$$

$$= (R_{12}R_{23}R_{31})\frac{\left(\sum R_{12}\right)}{\left(\sum R_{12}\right)^2}$$

$$= R_{12}\left[\frac{R_{23}R_{31}}{\sum R_{12}}\right]$$

By using R_3 from eq. (7),

$$R_1R_2 + R_2R_3 + R_3R_1 = R_{12}R_3$$

Hence,

$$R_{12} = \frac{R_1R_2 + R_2R_3 + R_3R_1}{R_3}$$

or,

$$R_{12} = R_1 + R_2 + \frac{R_1 R_2}{R_3} \qquad ...(11)$$

Similalary,

$$R_{23} = R_2 + R_3 + \frac{R_2 R_3}{R_1} \qquad ...(12)$$

and

$$R_{31} = R_3 + R_1 + \frac{R_3 R_1}{R_2} \qquad ...(13)$$

Thus equations (11), (12) and (13) will be used for star to delta conversion.

Delta (Δ) – Wye Conversion (Y)

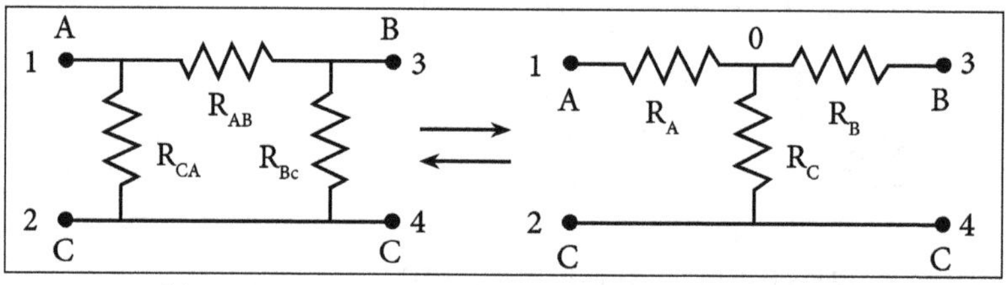

(a) (b)

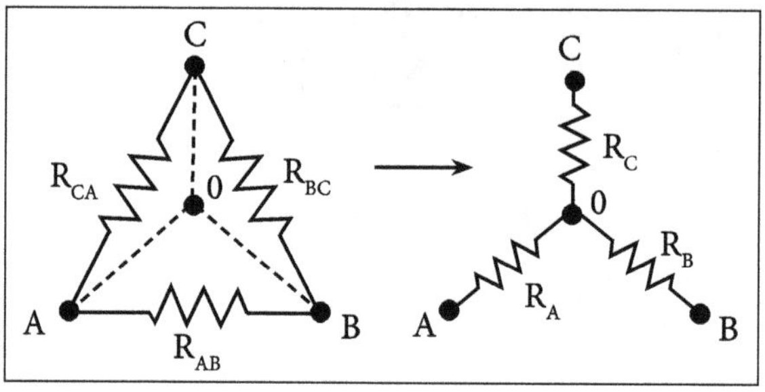

(c)

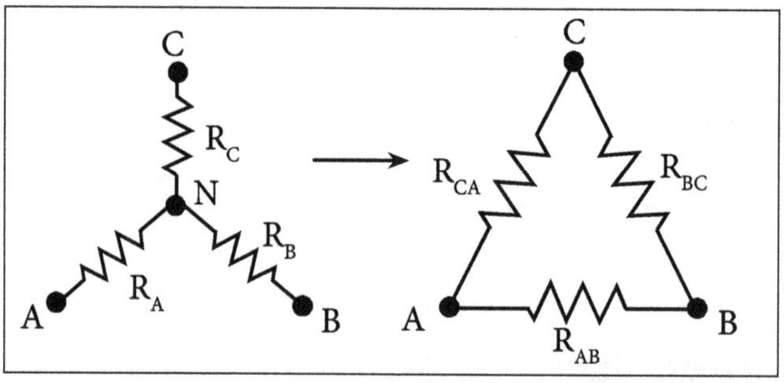

(d)

These configurations may be handled by the use of a Δ–Y or –ΔY transformation. One of the most basic three-terminal network equivalent is that of three resistors connected in "Delta (Δ)" and in "Wye(Y)". These two circuits identified in figure (c) and figure (d) are sometimes a part of the larger circuit and obtains their names from their configurations. These three terminal networks can be redrawn as four-terminal networks shown in figure (a) and figure (b).

We can obtain useful expression for direct transformation or conversion from Δ to Y or Y to Δ by considering that for equivalence, the two networks have the same resistance when looked at the similar pairs of terminals.

Conversion from Delta (Δ) to Star or Wye (Y)

Consider the network shown in figure (c) and assume the resistances (R_{AB}, R_{BC} and R_{CA}) in Δ network are known. Let us find the values of R_A, R_B and R_C in Wye (Y) network figure (a) that will produce the same resistance when measured between similar pairs of terminals. We can write the equivalence resistance between any two terminals in the following form.

Between A and C terminals:

$$R_A + R_C = \frac{R_{CA}(R_{AB} + R_{BC})}{R_{AB} + R_{BC} + R_{CA}} \qquad ...(1)$$

Between C and B terminals:

$$R_C + R_B = \frac{R_{BA}(R_{AB} + R_{CA})}{R_{AB} + R_{BC} + R_{CA}} \qquad ...(2)$$

Between B and A terminals:

$$R_B + R_A = \frac{R_{AB}(R_{CA} + R_{BC})}{R_{AB} + R_{BC} + R_{CA}} \qquad ...(3)$$

By combining above three equations, one can write an expression as given below.

$$R_A + R_B + R_C = \frac{R_{AB} R_{BC} + R_{BC} R_{CA} + R_{CA} R_{AB}}{R_{AB} + R_{BC} + R_{CA}} \qquad ...(4)$$

Subtracting equations (2), (1) and (3) from equation (4), we can write the express for unknown resistances of Wye (Y) network as,

$$R_A = \frac{R_{AB} R_{CA}}{R_{AB} + R_{BC} + R_{CA}} \qquad ...(5)$$

$$R_B = \frac{R_{AB}\,R_{BC}}{R_{AB}+R_{BC}+R_{CA}} \qquad ...(6)$$

$$R_C = \frac{R_{BC}\,R_{CA}}{R_{AB}+R_{BC}+R_{CA}} \qquad ...(7)$$

Problems

1. A 415V, 3-phase, 4 wire, star-connected system supplies three resistive loads as shown in figure below. Let us determine (a) the current in each line and (b) the current in the neutral conductor.

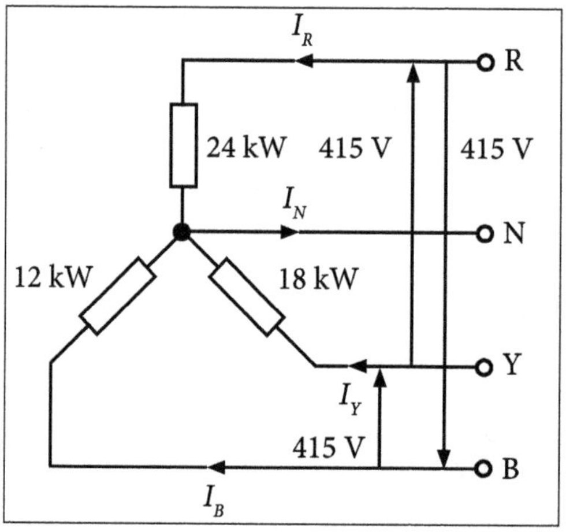

Solution:

Given:

A 415V, 3-phase, 4 wire, star-connected system supplies three resistive loads.

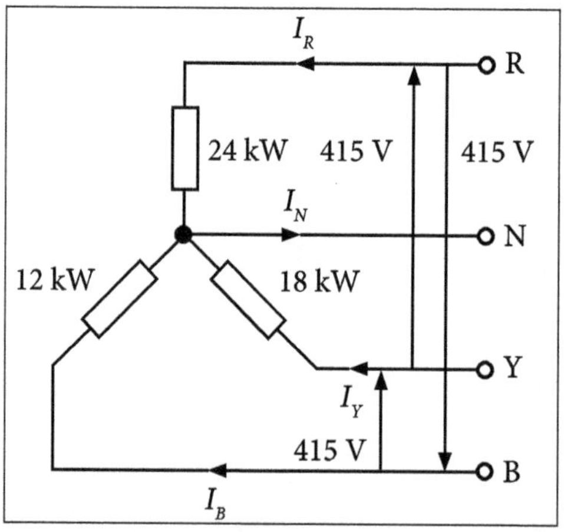

For a star-connected system $V_L = \sqrt{3}V_p$

Hence,

$$V_P \frac{V_L}{\sqrt{3}} = \frac{415}{\sqrt{3}} = 240$$

Since, current I = Power P/Voltage V for a resistive load.

Then,

$$I_R = P_R / V_R = 24\ 000 / 240 = 100A$$

$$I_Y = P_Y / V_Y = 18\ 000 / 240 = 75A$$

$$I_B = P_B / V_B = 12\ 000 / 240 = 50A$$

The three line currents are shown in the phasor diagram of figure. Since each load is resistive, the currents are in phase with the phase voltages and are mutually displaced by 1200.

The current in the neutral conductor is given by,

$$I_N = I_R + I_Y + I_B$$

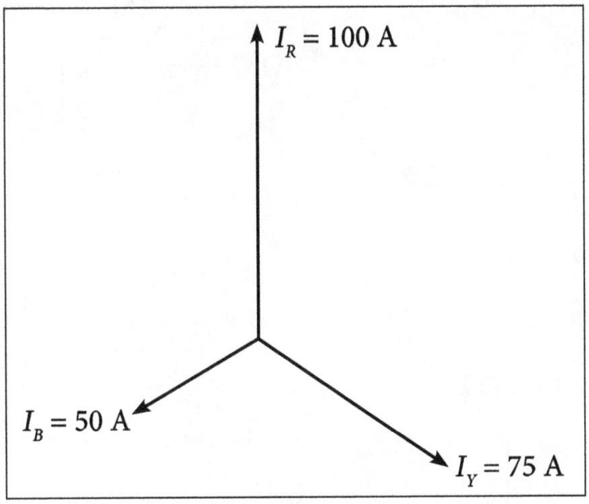

The above figure shows the three line currents added by phasor angle. O_a represents IR in magnitude and direction. From the nose of O_a, ab is drawn representing I_Y in magnitude and direction. From the nose of ab, bc is drawn representing I_B in its magnitude and direction O_c represents the resultant, I_N.

By measurement, $I_N = 43A$

Alternatively, by calculation, considering I_R at 90°, I_B at 210° and I_Y at 330°.

Total horizontal component $= 100 \cos 90° + 75 \cos 330° + 50 \cos 210° = 21.65$

Total vertical component $= 100 \sin 90° + 75 \sin 330° + 50 \sin 210° = 37.50$

Hence magnitude of $I_N = \sqrt{(21.65^2 + 37.50^2)} = 43.3A$

2. Let us convert the given delta figure into equivalent star.

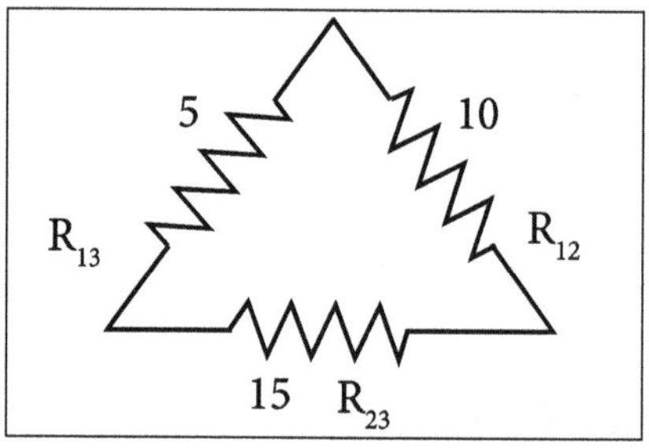

Solution:

Given:

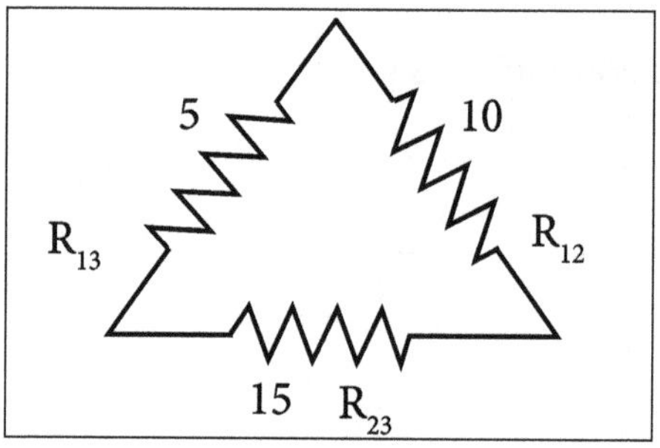

$R_1 = R_{12} \times R_{13} / R_{12} + R_{13} + R_{23}$

$R_1 = 10 \times 5 / 30 = 1.67\Omega$

$R_2 = 10 \times 15 / 30 = 5\Omega$

$R_3 = 5 \times 15 / 30 = 2.5\Omega$

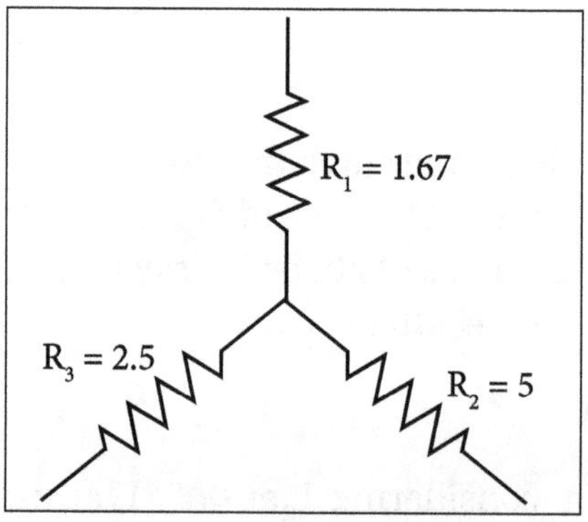

3

Coupled Circuits and Resonance

3.1 Coupled Circuits: Self-inductance, Mutual Inductance and Coefficient of Coupling

An electric circuit is said to be a coupled circuit, when there exists a mutual inductance between the coils (or inductors) present in that circuit. Coil is nothing but the series combination of resistor and inductor. In the absence of resistor, coil becomes inductor. Sometimes, the terms coil and inductor are interchangeably used.

Classification of Coupling

We can classify coupling into the following two categories:

- Electrical Coupling.
- Magnetic Coupling.

3.1.1 Self Inductance

Self-inductance of a coil is defined as the ratio of self-induced emf to the rate of change of current in the coil. It is denoted by 'L' and it depends upon the physical characteristics of the coil. Unit of self-inductance is Henry.

Let us consider a coil consists of N turns and carrying current I in the counter-clockwise direction, as shown in the below figure. If the current is steady, then the magnetic flux through the loop will remain constant. Suppose the current I changes with time, then according to Faraday's law, an induced emf will arise to oppose the change.

The induced current will flow clockwise if $dI/dt > 0$ and counterclockwise if $dI/dt < 0$. The property of the loop in which its own magnetic field opposes any change in current is termed as "self-inductance" and the emf generated is called the self-induced emf or back emf. All current-carrying loops exhibit this property. In particular, an inductor is a circuit element which has a large self-inductance.

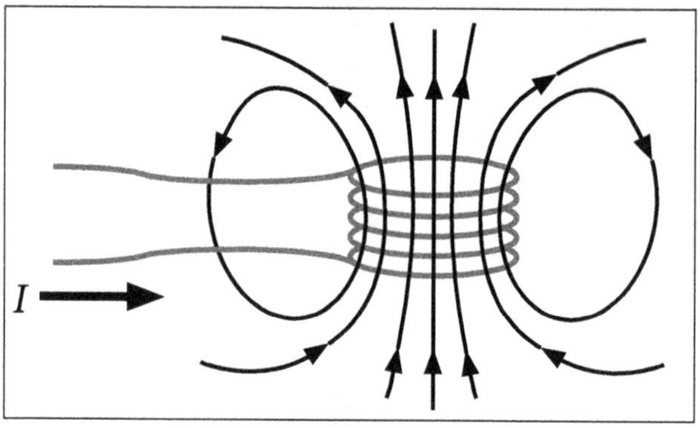

Magnetic flux through the current loop.

Mathematically, the self-induced emf can be written as,

$$\varepsilon_L = -N\frac{d\Phi_B}{dt} = -N\frac{d}{dt}\iint \vec{B}\cdot d\vec{A}$$

And is related to the self-inductance L by,

$$\varepsilon_L = -L\frac{dI}{dt}$$

The two expressions can be combined to yield,

$$L = \frac{N\Phi_B}{I}$$

Physically, the inductance L is a measure of an inductor's resistance to the change of current. Larger the value of L, lower the rate of change of current.

3.1.2 Mutual Inductance

Mutual inductance is a property associated with two or more coils or inductors which are in close presence of common magnetic flux which links the coils. A transformer is such a device whose operation is based on mutual inductance.

Let us consider two coils L_1 & L_2 which are close together, so that the flux produced by i_1 in coil L_1 also link coil L_2. When a voltage V_1 is applied across L_1, a current I_1 will start flowing through L_1, produces a flux **Φ**. This flux **Φ** also links coil L_2. If the current flowing through L_1 changes with respect to time, then the induced flux **Φ** also changes with respect to time.

The time varying flux surrounding the second coil L_2, induces an electromagnetic force (emf) or voltage across the coil L_2. This voltage is proportional to the rate of change of current flowing through coil L_1. Then, the two coils are said to be inductively coupled and is called mutual inductance.

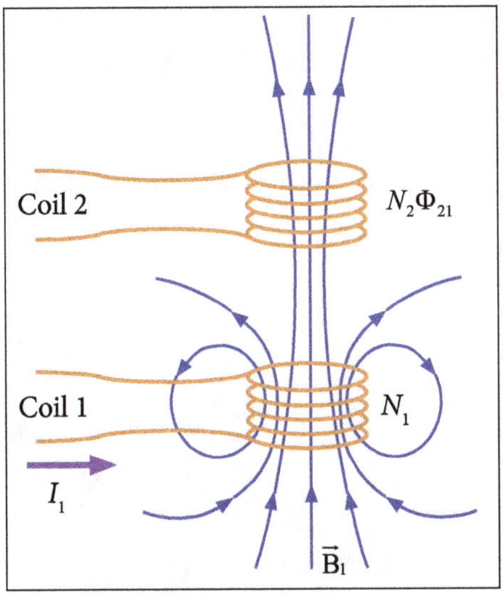

Changing current in coil 1 produces changing magnetic flux in coil 2.

Let the two coils are placed near to each other, as shown in the above figure. The first coil has N_1 turns and carries a current I_1 which gives rise to a magnetic field B_1. Since the two coils are close to each other, some of the magnetic field lines through coil 1 will also pass through coil 2.

Let Φ_{21} denote the magnetic flux through one turn of coil 2 due to I_1. Now, by varying I_1 with time, there will be an induced emf associated with the changing magnetic flux in the second coil.

i.e., $$\varepsilon_{21} = -N_2 \frac{d\Phi_{21}}{dt} = -\frac{d}{dt} \iint_{\text{coil } 2} \vec{B_1} \cdot d\vec{A_2}$$

The time rate of change of magnetic flux Φ_{21} in coil 2 is proportional to the time rate of change of the current in coil 1.

i.e., $$N_2 = \frac{d\Phi_{21}}{dt} = M_{21} \frac{dI_1}{dt}$$

Where, the proportionality constant M_{21} is called the mutual inductance.

It can also be written as,

$$M_{21} = \frac{N_2 \Phi_{21}}{I_1}$$

The SI unit for inductance is henry (H).

$$1 \text{ henry} = 1 \text{ H} = 1 \text{ T m}^2/\text{A}$$

The mutual inductance M_{21} depends only on the geometrical properties of the two coils such as the number of turns and the radii of the two coils. In a similar manner, instead there is a current I_2 in the second coil and it is varying with time, then the induced emf in coil 1 becomes,

$$\varepsilon_{12} = -N_1 \frac{d\Phi_{12}}{dt} = -\frac{d}{dt} \iint_{coil\ 1} \vec{B_2} \cdot d\vec{A_1}$$ and a current is induced in coil 1.

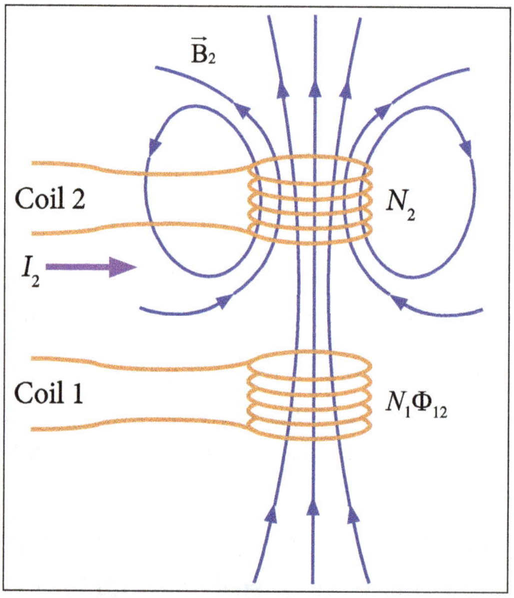

Changing current in coil 2 produces changing magnetic flux in coil 1.

This changing flux in coil 1 is proportional to the changing current in coil 2,

$$N_1 = \frac{d\Phi_{12}}{dt} = M_{12} \frac{dI_2}{dt}$$

Where, the proportionality constant M_{12} is another mutual inductance and can be written as,

$$M_{12} = \frac{N_1 \Phi_{12}}{I_2}$$

However, using the reciprocity theorem which combines the Ampere's law and the Biot- Savart law, we may show that the constants are equal,

$$M_{12} = M_{21} \equiv M$$

3.1.3 Coefficient of Coupling

In coupled coils, the coefficient of coupling is defined as the fraction of the total flux produced by one coil linking another coil.

The amount of coupling between the inductively coupled coils is expressed in terms of the coefficient of coupling, which is defined as,

$$K=\frac{\Phi_{12}}{\Phi_1}=\frac{\Phi_{21}}{\Phi_2}$$

[At K = 1, maximum mutual inductance occurs in coupled circuits]

Where,

M - Mutual inductance between the two coils.

L_1 – Self-inductance of the first coil.

L_2 – Self-inductance of the second coil.

3.2 Analysis of Coupled Circuits and Natural Current

Inductively coupled multi-mesh circuits can be analysed using Kirchhoff's laws and by loop current methods. Let us consider the below figure, where three coils are inductively coupled. For such a system of inductors, we can define a inductance matrix L as,

$$L=\begin{bmatrix} L_{11} & L_{12} & L_{13} \\ L_{21} & L_{22} & L_{23} \\ L_{31} & L_{32} & L_{33} \end{bmatrix}$$

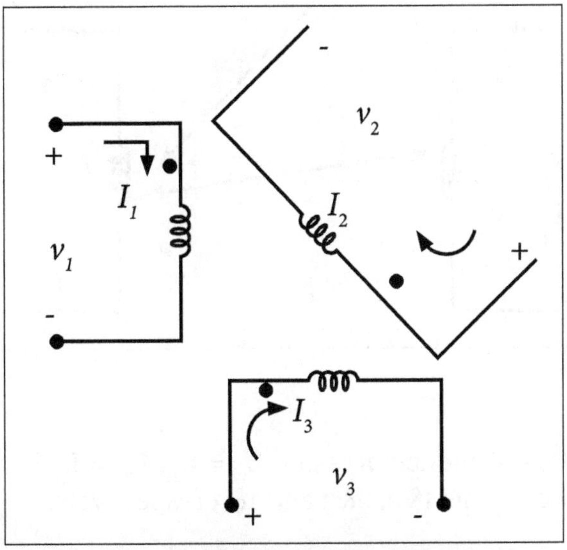

Inductively coupled circuits.

Where L_{11}, L_{22} and L_{33} are self-inductances of the coupled circuits and L_{12} = L_{21}, L_{23} =

L_{32} and $L_{13} = L_{31}$ are mutual inductances. More precisely, L_{12} is the mutual inductance between coils 1 and 2, L_{13} is the mutual inductance between coils 1 and 3 and L_{23} is the mutual inductance between coils 2 and 3. The inductance matrix has its order equal to the number of inductors and is symmetric. In terms of voltages across the coils, we have a voltage vector related to i by,

$$[v]=[L]\left[\frac{di}{dt}\right]$$

Where, v and i are the vectors of the branch voltages and currents, respectively. Thus, the branch volt-ampere relationships of the three inductors are given by,

$$\begin{bmatrix} v_1 \\ v_2 \\ v_3 \end{bmatrix} = \begin{bmatrix} L_{11} & L_{12} & L_{13} \\ L_{21} & L_{22} & L_{23} \\ L_{31} & L_{32} & L_{33} \end{bmatrix} \begin{bmatrix} di_1/dt \\ di_2/dt \\ di_3/dt \end{bmatrix}$$

Using KVL and KCL, the effective inductances can be calculated. The polarity for the inductances can be determined by using passivity criteria, whereas the signs of the mutual inductances can be determined by using the dot convention.

Example 1: For the circuit shown in the below figure, write the inductance matrix.

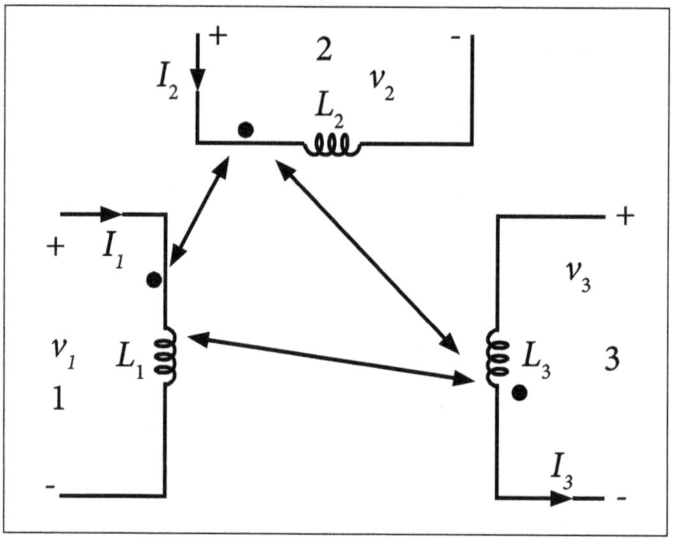

Solution:

Let L_1, L_2 and L_3 be the self-inductances and $L_{12} = L_{21}$, $L_{23} = L_{32}$ and $L_{13} = L_{31}$ be the mutual inductances between coils 1&2, 2&3 and 1&3 respectively.

$L_{12} = L_{21}$ is positive, as both the currents are entering at dot marked terminals, whereas $L_{13} = L_{31}$ and $L_{23} = L_{32}$ are negative.

Thus, the inductance matrix is given as,

$$L=\begin{bmatrix} L_1 & L_{12} & -L_{13} \\ L_{21} & L_2 & -L_{23} \\ -L_{31} & -L_{32} & L_3 \end{bmatrix}$$

3.2.1 Nature Current

Further understanding of coupled coils is achieved from consideration of a passive second loop as shown in the below figure. Source v_1 drives a current i_1, with a corresponding flux $Ø_1$ as shown. Now Lenz's law implies that the polarity of the induced voltage in the second circuit is such that if the circuit is completed, a current will pass through the second coil in such a direction as to create a flux opposing the main flux established by i_1. That is, when the switch is closed in the below figure, flux $Ø_2$ will have the direction shown.

The right-hand rule, with the thumb pointing in the direction of $Ø_2$, provides the direction of the natural current i_2. The induced voltage is the driving voltage for the second circuit, as suggested in the below figure. This voltage is present whether or not the circuit is closed. When the switch is closed, current i_2 is established, with a positive direction as shown.

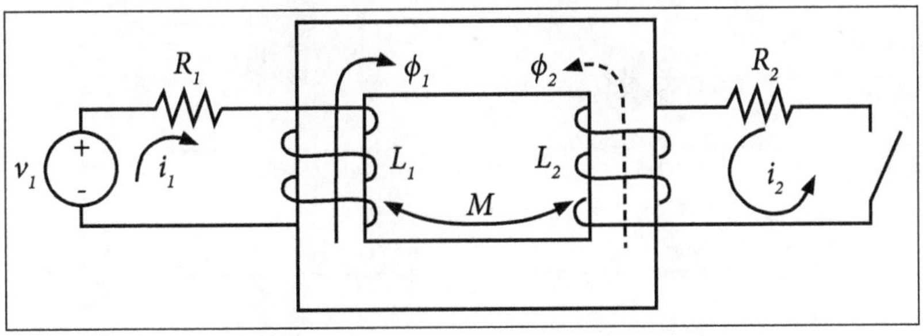

3.3 Dot Rule of Coupled Circuits and Conductively Coupled Equivalent Circuits

In circuit analysis, the dot convention is the convention used to denote the voltage polarity of mutual inductance of two component. The two conventions are:

- If a current "enters" the dotted terminal of a coil, the reference polarity of the voltage induced in the other coil is positive at its dotted terminal.
- If a current "leaves" the dotted terminal of a coil, the reference polarity of the voltage induced in the other coil is negative at its dotted terminal.

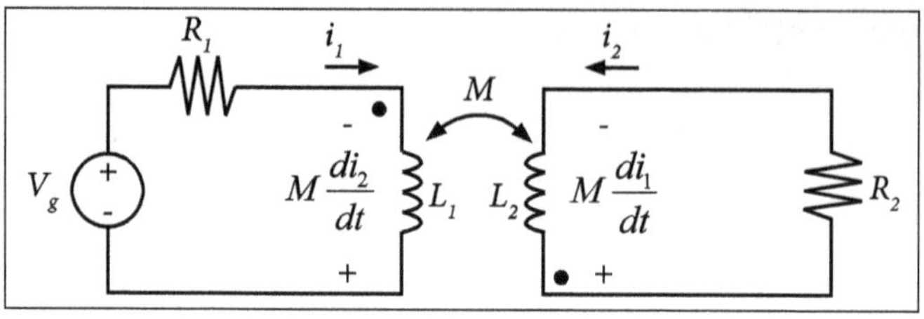

Dot convention.

The equivalent circuits of all possible dot convention are shown in the figure below.

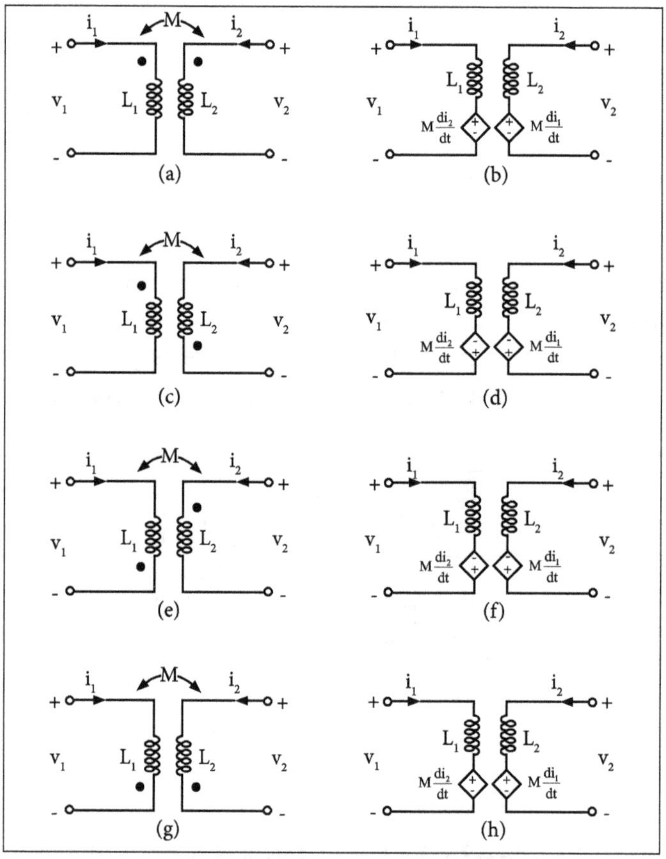

Equivalent circuits of dot convention.

Let us consider a magnetically coupled circuit with dots placed as shown in the figure (a). Both the currents, i_1 and i_2 are entering the dotted terminals. Hence according to the dot convention, the mutually induced e.m.f. in both the coils has the polarity same as self-induced e.m.f. in respective coil. The equivalent circuit is as shown in the above figure (b).

By applying KVL, the network equations of the equivalent circuit can be written as,

$$v_1 = L_1 \frac{di_1}{dt} + M\frac{di_2}{dt}$$

$$v_2 = L_2 \frac{di_2}{dt} + M\frac{di_1}{dt} \quad \text{...(1)}$$

Let us consider magnetically coupled circuit as shown in the figure (c) with dot placed at lower terminal of coil L_2. Hence current i_1 enters through dotted terminal of L_1 while current i_2 leaves through dotted terminal of L_2.

So according to dot convention, the polarity of mutually induced e.m.f. in L_1 due to i_2 in L_2 will be opposite to that of self-induced e.m.f. in coil L. Also the polarity of mutually induced e.m.f. in coil L_2 due to the current i_1 in coil L_1 will be opposite to that of self-induced e.m.f. in coil L_2. The equivalent circuit is as shown in the figure (d).

By using KVL, the network equations can be written as,

$$v_1 = L_1 \frac{di_1}{dt} - M\frac{di_2}{dt}$$

$$v_2 = L_2 \frac{di_2}{dt} - M\frac{di_1}{dt} \quad \text{...(2)}$$

For the equivalent circuit shown in the figure (f).

By applying KVL, the network equations can be written as,

$$v_1 = L_1 \frac{di_1}{dt} - M\frac{di_2}{dt}$$

$$v_2 = L_2 \frac{di_2}{dt} - M\frac{di_1}{dt} \quad \text{...(3)}$$

For last possible combination, both the dots are placed at lower terminals of coils L_1 and L_2. Also both the currents leave dot as shown in the figure (g). The equivalent circuit is as shown in the above figure (h).

By applying KVL, the network equations can be written as,

$$v_1 = L_1 \frac{di_1}{dt} + M\frac{di_2}{dt}$$

$$v_2 = L_2 \frac{di_2}{dt} + M\frac{di_1}{dt} \quad \text{...(4)}$$

The analysis of multi-winding inductor networks can be carried out for each pair of windings using same dot convention.

Problem

1. An iron ring of relative permeability 100 is wound uniformly with two coils of 100 and 400 turns of wire. The cross-section of the ring is 4cm². The mean circumference is 50cm. Let us calculate,

1. The self-inductance of each of the two coils.
2. The mutual inductance.
3. Total inductance when the coils are connected in series with the flux in the same sense.
4. Total inductance when the coils are connected in series with flux in the opposite sense.

Solution:

Given:

Relative permeability = 100

Number of turns of wire = 100 and 400

Cross-section of the ring = 4cm²

Mean circumference = 50cm

Formula to be used:

$$L=\frac{\mu_0 N^2 A}{2\pi R_m}$$

$$M=\frac{\mu_0 N_1 N_2 A}{2\pi R_m}$$

Total inductance = $L_1 + L_2 + 2M$

(1) Self-inductance of coil with 100 turns is given by,

$$L=\frac{\mu_0 N^2 A}{2\pi R_m}=\frac{4\pi\times10^{-7}\times100^2\times4}{50\times10^{-2}}$$

$= 0.1$ Henry

Self-inductance with 400 turns is given by,

$$L=\frac{4\pi\times10^{-7}\times400^2\times4}{50\times10^{-2}}=1.61\,\text{Henry}$$

(2) Mutual inductance of the coil is given by,

$$M=\frac{\mu_0 N_1 N_2 A}{2\pi R_m}$$

$$=\frac{4\pi\times10^{-7}\times100\times400\times4}{50\times10^{-2}}$$

$$=0.402 \text{ Henry}$$

(3) Total inductance $= L_1 + L_2 + 2M$

$$=0.1+1.61+2\times0.4$$

$$=2.51$$

(4) Total inductance when the coils are in series with flux in the opposite sense is given by,

$$L_1 + L_2 - 2M$$

$$=0.1+1.61-2\times0.4$$

$$=0.91$$

3.3.1 Conductively Coupled Equivalent Circuits

From the point of view of electrical circuit analysis, usually it is desirable to replace magnetically coupled circuit with an equivalent network called conductively coupled circuit.

In this circuit, no magnetic couplings are involved. The dot convention is also not needed in conductively coupled circuit and can be analyzed by general network simplification techniques such as mesh analysis, node analysis etc.

Consider magnetically coupled circuit and its equivalent T section as shown in the figure below.

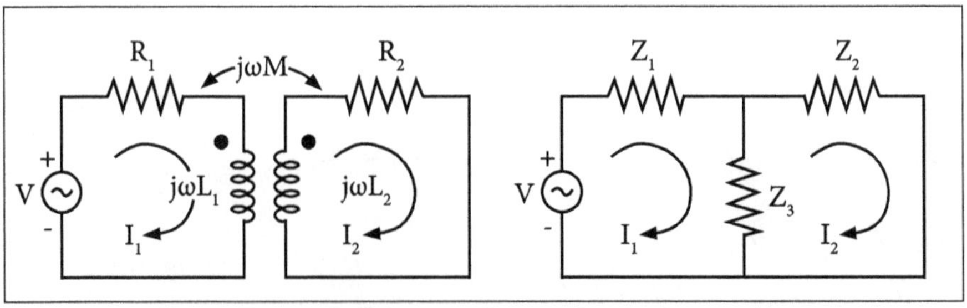

Magnetically coupled circuit and equivalent T section of circuit.

Consider the magnetically coupled circuit,

By applying KVL to loop 1, we get,

$$-(R_1)I_1-(j\omega L_1)I_1+(j\omega M)I_2+V=0$$

$$\therefore \quad -[R_1+j\omega L_1]I_1+[j\omega M]I_2=-V$$

$$(R_1+j\omega L_1)I_1-(j\omega M)I_2=V \qquad ...(1)$$

By applying KVL to loop 2, we get,

$$+(j\omega M)I_1-(R_2)I_2-(j\omega L_2)I_2=0$$

$$-(j\omega M)I_1+(R_2+j\omega L_2)I_2=0 \qquad ...(2)$$

Writing equations (1) and (2) in matrix form, we obtain as,

$$\begin{bmatrix}(R_1+j\omega L_1) & -j\omega M\\ -j\omega M & (R_2+j\omega L_2)\end{bmatrix}\begin{bmatrix}I_1\\ I_2\end{bmatrix}=\begin{bmatrix}V\\ 0\end{bmatrix} \qquad ...(3)$$

Consider the equivalent T section of circuit,

Applying KVL to loop 1, we get,

$$-(Z_1)I_1-(Z_3)I_1+(Z_3)I_2+V=0$$

$$\therefore \quad -(Z_1+Z_3)I_3+(Z_3)I_2=-V$$

$$\therefore \quad (Z_1+Z_3)I_1-(Z_3)I_2=V \qquad ...(4)$$

Applying KVL to loop 2, we get,

$$-(Z_2)I_2-(Z_3)I_2+(Z_3)I_1=0$$

$$(Z_3)I_1-(Z_2+Z_3)I_2=0$$

$$-(Z_3)I_1+(Z_2+Z_3)I_2=0 \qquad ...(5)$$

Writing equations (4) and (5) in matrix form, we obtain as,

$$\begin{bmatrix}(Z_1+Z_3) & -Z_3\\ -Z_3 & (Z_2+Z_3)\end{bmatrix}\begin{bmatrix}I_1\\ I_2\end{bmatrix}=\begin{bmatrix}V\\ 0\end{bmatrix} \qquad ...(6)$$

To have the circuit shown in figure above, impedance matrix in equation (3) and (6) must be equal,

$$Z_1 + Z_3 = R_1 + j\omega L_1$$

$$Z_2 + Z_3 = R_2 + j\omega L_2$$

$$Z_3 = j\omega M \qquad \text{...(7)}$$

From the above equation, we can write as,

$$Z_1 = R_1 + j\omega L_1 - j\omega M = R_1 + j\omega(L_1 - M)$$

$$Z_2 = R_{21} + j\omega L_2 - j\omega M = R_2 + j\omega(L_2 - M)$$

$$Z_3 = j\omega M$$

The conductively coupled equivalent circuit of magnetically coupled circuit is shown in figure as,

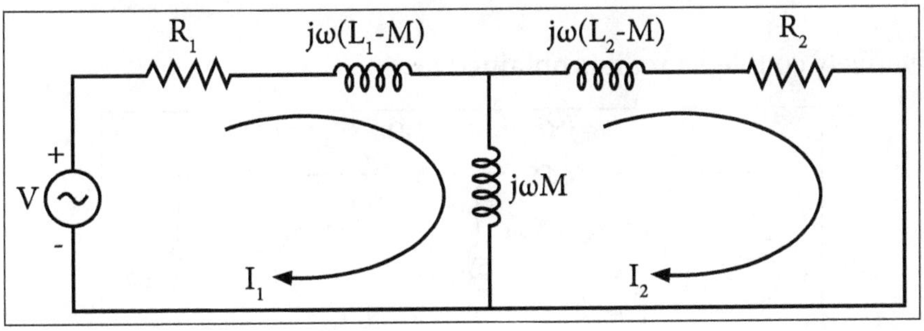

Problems

1. Let us determine the conductively coupled circuit for the figure shown below.

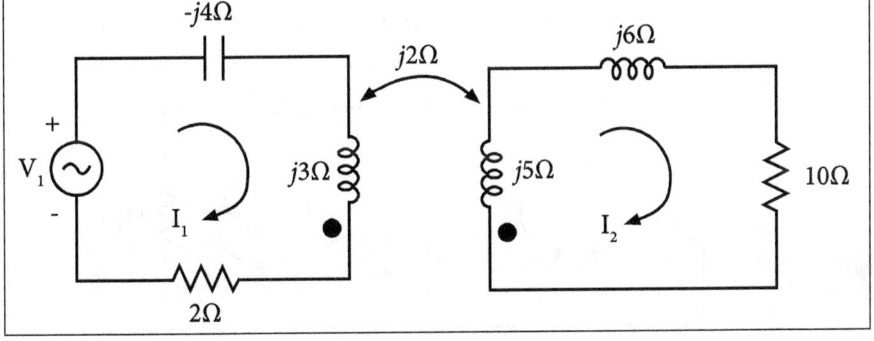

Solution:

Given:

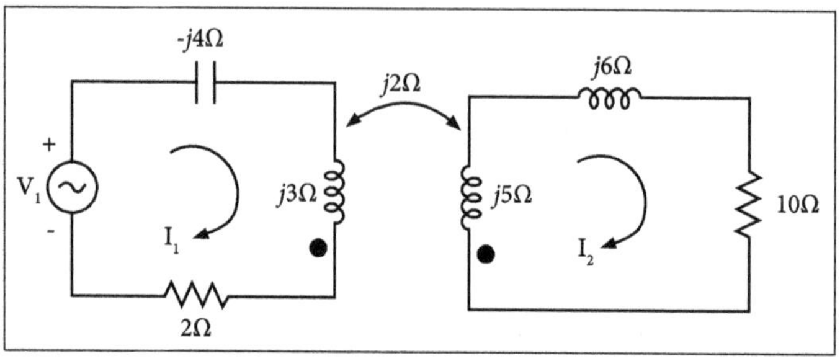

The current I_1 leaves from the dotted end and enters I_2. Hence the mutual inductance M is negative.

In the conductively coupled circuits,

$$Z_1 = j\omega(L_1 - M) = j\omega L_1 - j\omega M = j3 - j2 = j1\Omega$$

$$Z_2 = j\omega(L_2 - M) = j\omega L_2 - j\omega M = j5 - j2 = j3\Omega$$

$$Z_3 = j\omega M = j2\Omega$$

The conductively coupled circuit is obtained as,

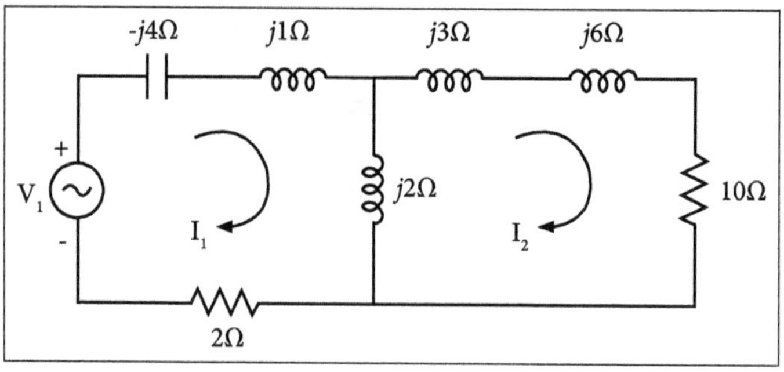

2. Let us draw the conductively coupled circuit for the figure shown below.

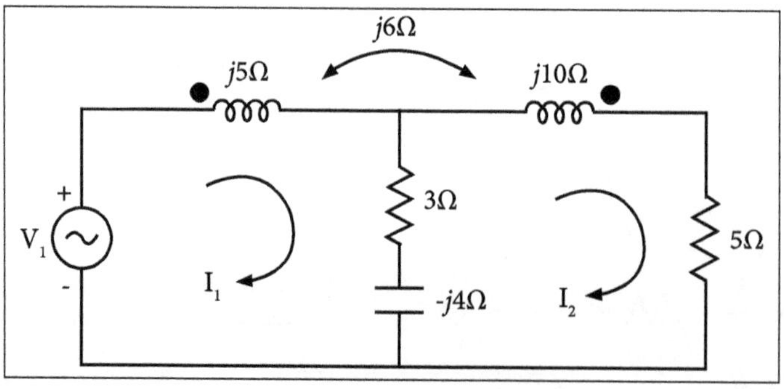

Solution:

Given:

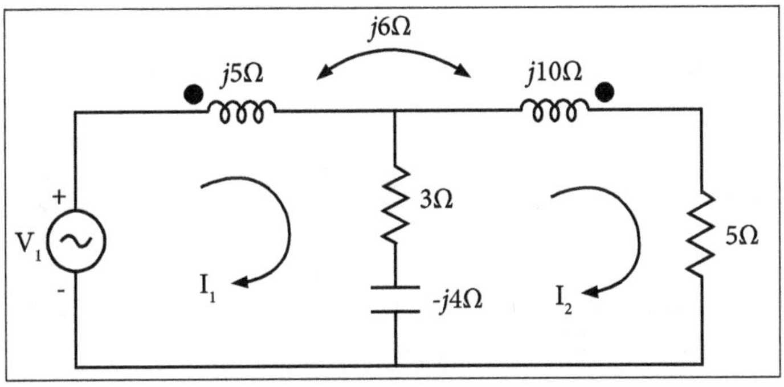

The current I_1 enters from the dotted end and leaves the I_2. Hence, the mutual inductance M is negative.

In the conductively coupled circuits,

$$Z_1 = j\omega(L_1 - M) = j\omega L_1 - j\omega M = j5 - j6 = j1\ \Omega$$

$$Z_2 = j\omega(L_2 - M) = j\omega L_2 - j\omega M = j10 - j6 = j4\ \Omega$$

$$Z_3 = j\omega M = j6\Omega$$

The conductively coupled circuit is obtained as,

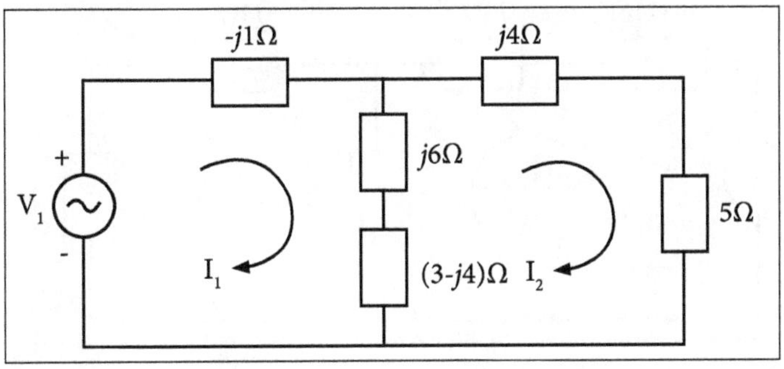

3. Let us determine the conductively coupled circuit for the network shown below.

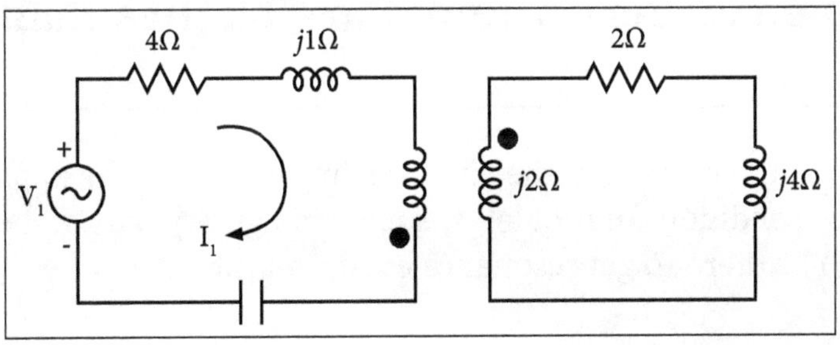

Solution:

Given:

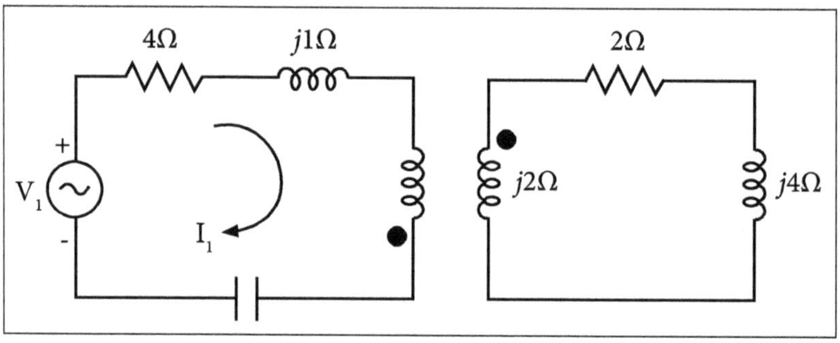

The current I_1 and I_2 leave from the dotted terminal. Hence the mutual inductance is positive.

In the conductively coupled circuits,

$$Z_1 = j\omega(L_1 + M) = j\omega L_1 + j\omega M = j4 + j2 = j6\,\Omega$$

$$Z_2 = j\omega(L_2 + M) = j\omega L_2 + j\omega M = j2 + j2 = j4\,\Omega$$

$$Z_3 = j\omega M = -j2\,\Omega$$

The conductively coupled circuit is obtained as,

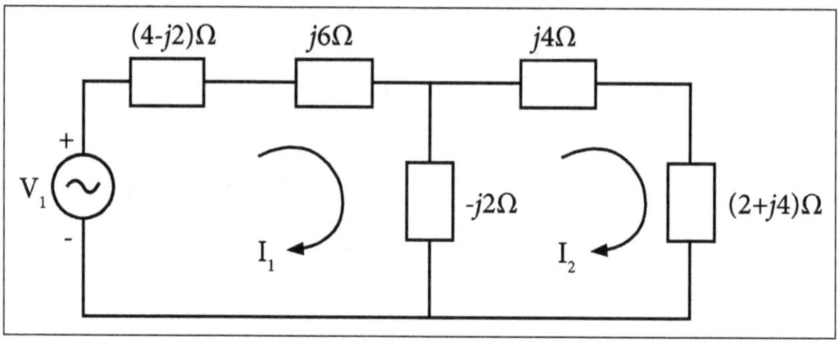

3.4 Resonance: Series and Parallel Resonance and its Bandwidth

When $X_L = X_C$ [or] the current in the circuit is in-phase with the applied voltage, it is called Resonance condition. In problems, equate imaginary part of the Impedance (Z) or Admittance (Y) to zero to get resonance condition [i.e., if $Z = Re + jI_m$, then equate $I_m = 0$].

Two types of Resonance circuits:

- Series Resonance.
- Parallel Resonance.

3.4.1 Series Resonance and Bandwidth of Series Resonance

The basic series-resonant circuit is shown in the below figure. As the frequency of the source changes, the maximum amplitude of the source voltage (V_m) is held constant.

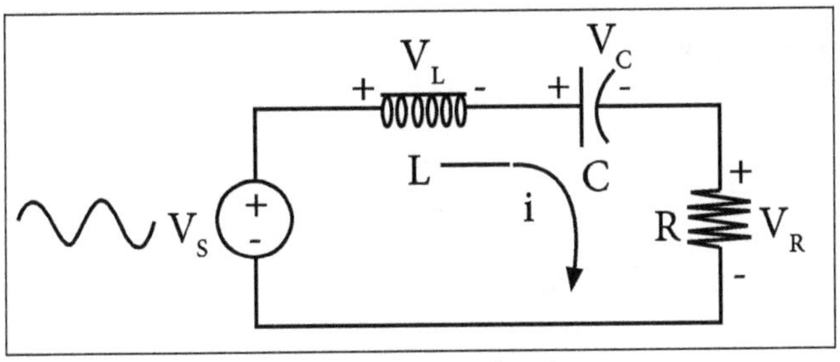

Series resonant circuit.

From the above figure,

$$V_S = V_m \cos(\omega t)$$

$$i = I_m \cos(\omega t + \theta)$$

The frequency at which the reactance of the inductance and the capacitance cancel each other is the resonant frequency of this circuit.

This occurs at,

$$\omega_o = 1/\sqrt{(LC)}$$

Since $i = V_R/R$, then the current i can be studied by the voltage across the resistor.

The current i has the expression,

$$i = I_m \cos(\omega t + \theta)$$

$$I_m = \frac{V_m}{\sqrt{R^2 + \left(\omega L - \frac{1}{\omega C}\right)^2}}$$

$$\theta = \tan^{-1}\left(\frac{\omega L - \frac{1}{\omega C}}{R}\right)$$

Bandwidth of Series Resonance

$$B = \omega_2 - \omega_1 = R/L$$

$$\omega_{2,1} = \sqrt{\left(\frac{R}{2L}\right)^2 + \frac{1}{LC}} \pm \frac{R}{2L}$$

$\omega_{2,1}$ is the half power frequency or the 3 dB frequency, (i.e.,) the frequencies at which the value of I_m equals the maximum possible value divided by 1.414.

The quality factor is given as,

$$Q = \frac{\omega_0}{B} = \frac{1}{R}\sqrt{\frac{L}{C}}$$

Then the maximum value of V_L occurs at,

$$\omega = \frac{\omega_0}{\sqrt{1 - \frac{R^2C}{2L}}}$$

V_C occurs at,

$$\omega_0 \sqrt{1 - \frac{R^2C}{2L}}$$

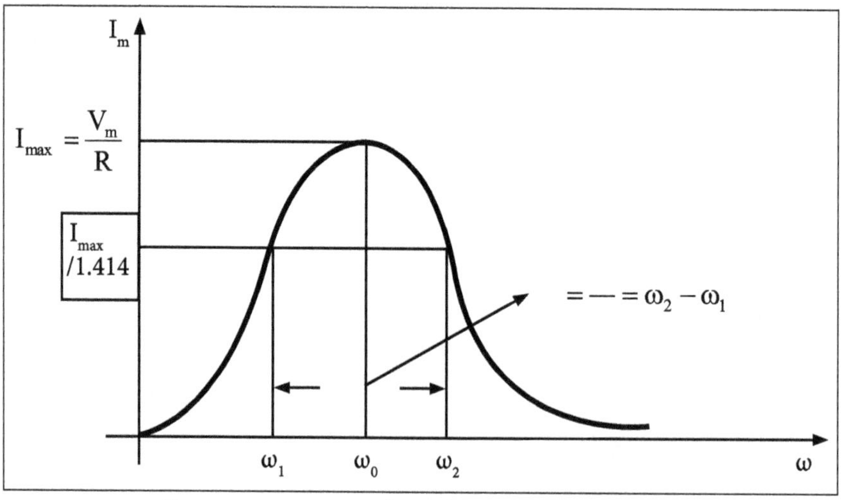

Frequency response of a Series - Resonant Circuit.

Quality Factor, Q

Reactive components such as capacitors and inductors are often described with a figure of merit called Q. While it can be defined in many ways, but the most fundamental description is,

$$Q=\omega \frac{\text{energy stored}}{\text{average power dissipated}}$$

In other words,

The quality factor Q, is the ratio of the reactive power in the inductor or capacitor to the true power in the resistance in series with the coil or capacitor,

$$Q=2\pi\times \frac{\text{max energy stored}}{\text{energy dissipated per cycle}}$$

Quality factor of the capacitance circuit $Q=\frac{1}{wCR}$

Thus, it is a measure of the ratio of stored lost energy per unit time. Note that this definition does not specify what type of system is required. Thus, it is quite general.

An ideal reactive component (capacitor or inductor) stores energy,

$$E \ \frac{1}{2}CV_{pk}^2 \quad \text{or} \quad \frac{1}{2}LI_{pk}^2$$

Since any real component also loss due to the resistive component, the average power dissipated is,

$$P_{avg} = \frac{}{2} I_{pk} R = \frac{pk}{2R}$$

Here, we consider an example of a series resonant circuit.

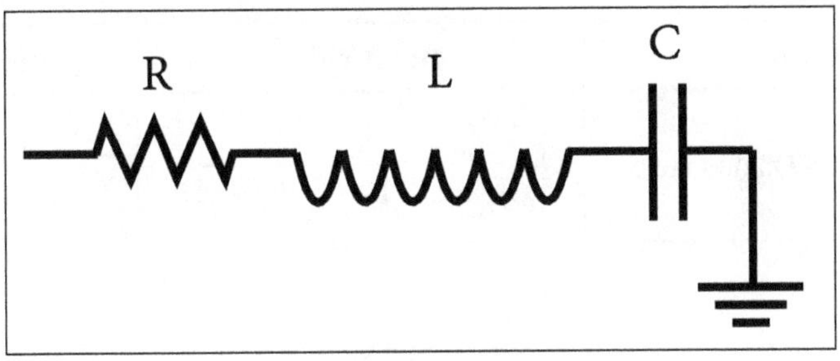

Series - Resonant circuit.

At resonance, the reactances cancel out leaving just a peak voltage, V_{pk} across the loss resistance, R. Thus, $I_{pk} = V_{pk} / R$ is the maximum current which passes through all elements.

Then,

$$Q=\omega_0 \frac{LI_{pk}^2/2}{I_{pk}^2 R/2}=\frac{\omega_0 L}{R}=\frac{1}{\omega_0 RC}$$

$$Q=\frac{1}{\omega RC}=\frac{X}{R}$$

Where we pretend that the capacitor is resonated with an ideal inductor at frequency ω. X is the capacitive reactance and R is the series resistance. Since this Q refers only to the capacitor itself, in isolation from the rest of the circuit, it is called unloaded Q or Q_U. The higher the unloaded Q, the lower the loss. Notice that the Q decreases with frequency.

The unloaded Q of an inductor is given by,

$$Q_U=\frac{\omega_0 L}{R}$$

Where, R is a series resistance.

Note that Q is proportional to frequency for an inductor.

The value of Q of an inductor will depend upon the wire diameter, core material and whether it is in a shielded metal or not. It is easy to show that for a parallel resonant circuit, the Q is given by susceptance/conductance.

$$Q=\frac{B}{G}$$

Where,

B is the susceptance of the capacitor or inductor and G is the shunt conductance.

Quality Factor for Series RLC Circuit

Circuit	Bandwidth	Capacitive	Inductive
Series RLC circuit- Quality factor	$Q=\frac{f_r}{f_2-f_1}=\frac{f_r}{BW}=\frac{f_r}{R/2\pi L}$	$Q=\frac{1}{wRC}$	$Q=\frac{wL}{R}$

Bandwidth

At a certain frequency, the power dissipated by the resistor is half of the maximum power which occurs at,

$$\omega_0=\frac{1}{\sqrt{LC}}$$

The half power occurs at the frequencies for which the amplitude of the voltage across the resistor becomes equal to $1/\sqrt{2}$ of the maximum.

$$P_{1/2} = \frac{1}{4}\frac{V_{max}^2}{R}$$

The below figure shows the graphical form of the various frequencies of interest.

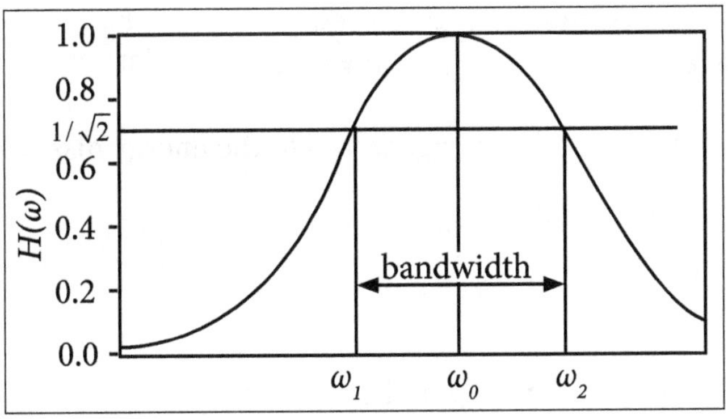

Graphical representation of the various frequencies of interest.

Therefore, the ½ power occurs at the frequencies for which,

$$\frac{1}{\sqrt{2}} = \frac{\omega RC}{\sqrt{\left(1-\omega^2 LC\right)^2 + \left(\omega RC\right)^2}}$$

Equation has two roots,

$$\omega_1 = -\frac{R}{2L} + \sqrt{\left(\frac{R}{2L}\right)^2 + \frac{1}{\omega_0^2}}$$

$$\omega_2 = -\frac{R}{2L} + \sqrt{\left(\frac{R}{2L}\right)^2 + \frac{1}{\omega_0^2}}$$

$$\text{Bandwidth} = B = \omega_2 - \omega_1$$

By multiplying these two equations we can show that ω_o is the geometric mean of ω_1 and ω_2,

$$\omega_o = \sqrt{\omega_1 \omega_2}$$

From the plot on the graph, the bandwidth increases with increasing R. Equivalently, the sharpness of the resonance increases with decreasing R.

For a fixed L and C, a decrease in R corresponds to a narrower resonance and thus a higher selectivity regarding the frequency range that can be passed by the circuit.

As we increase R, the frequency range over which the dissipative characteristics dominate the behavior of the circuit increases. In order to quantify this behavior, we define a parameter called Quality Factor Q which is related to the sharpness of the peak and it is given by,

$$Q=2\pi\frac{\text{maximum energy stored}}{\text{total energy lost per cycle at resonance}}=2\pi\frac{E_S}{E_D}$$

Which represents the ratio of the energy stored to the energy dissipated in a circuit.

The energy stored in the circuit is,

$$E_S=\frac{1}{2}LI^2+\frac{1}{2}CV_C^2$$

For $V_c = A\sin\omega t$, the current flowing in the circuit is,

$$I=C\frac{dV_C}{dt}\omega CA\cos(\omega t).$$

The total energy stored in the reactive elements is,

$$E_S=\frac{1}{2}L\omega^2C^2A^2\cos^2(\omega t)+\frac{1}{2}CA^2\sin^2(\omega t)$$

At the resonance frequency where,

$$\omega=\omega_0=\frac{1}{\sqrt{LC}}$$

The energy stored in the circuit becomes,

$$E_S=\frac{1}{2}CA^2$$

The energy dissipated per period is equal to the average resistive power dissipated times the oscillation period i.e.,

$$E_D=R\left\langle I^2\right\rangle\frac{2\pi}{\omega_0}=R\left(\frac{\omega_0^2C^2A^2}{2}\right)\frac{2\pi}{\omega_0}=2\pi\left(\frac{1}{2}\frac{RC}{\omega_0L}A^2\right)$$

And so, the ratio Q becomes,

$$Q=\omega_0L/R$$

By combining these equations, we obtain the relationship between the bandwidth and the Q factor.

Selectivity

The ratio of bandwidth to the resonant frequency is called selectivity.

That is,

$$S = \frac{B.W}{f_r}$$

Properties of RLC Resonant Circuits

Series RLC circuit - Resonance Frequency (f_r)

	Series	Parallel
Circuit	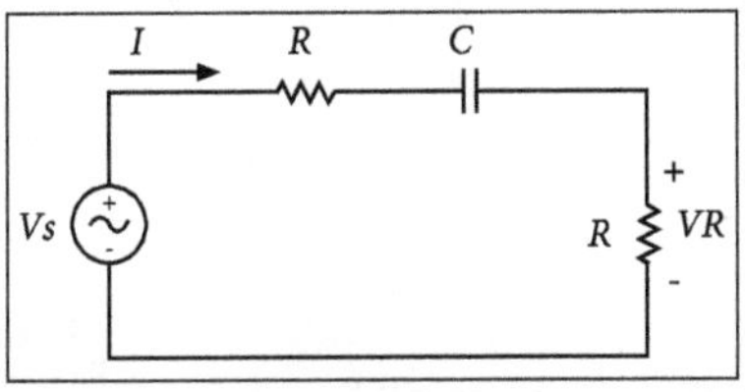	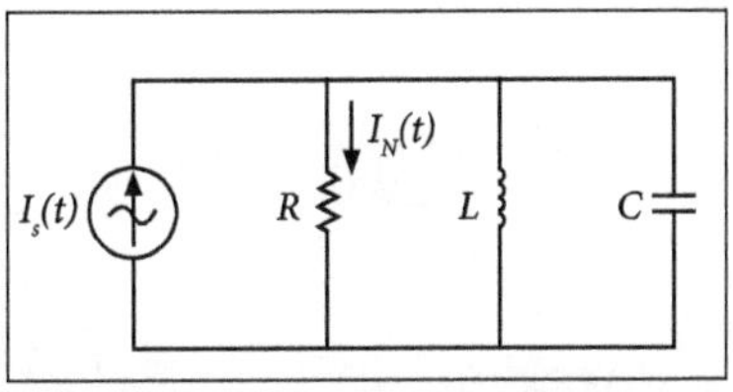
Transfer Function	$\lvert H(\omega)\rvert \equiv \left\lvert\frac{VR}{Vs}\right\rvert = \frac{\omega RC}{\sqrt{(1-\omega^2 LC)^2 + (\omega RC)^2}}$	$\lvert H(\omega)\rvert \equiv \left\lvert\frac{I_R}{I_S}\right\rvert = \frac{\omega L}{\sqrt{(R-\omega^2 LR)^2 + (\omega L)^2}}$
Resonant Frequency	$\omega_o = \frac{1}{\sqrt{LC}}$	$\omega_o = \frac{1}{\sqrt{LC}}$
½ power frequency	$\omega_1 = \frac{R}{2L} + \sqrt{\left(\frac{R}{2L}\right)^2 + \frac{1}{\omega_o^2}}$ $\omega_2 = \frac{R}{2L} + \sqrt{\left(\frac{R}{2L}\right)^2 + \frac{1}{\omega_o^2}}$	$\omega_1 = -\frac{1}{2RC} + \sqrt{\left(\frac{1}{2RC}\right)^2 + \frac{1}{\omega_o^2}}$ $\omega_2 = \frac{1}{2RC} + \sqrt{\left(\frac{1}{2RC}\right)^2 + \frac{1}{\omega_o^2}}$
Bandwidth	$B_S = \omega_2 - \omega_1 = \frac{R}{L}$	$B_P = \omega_2 - \omega_1 = \frac{1}{RC}$
Q factor	$Q = \frac{\omega_o}{B_S} = \frac{\omega_o L}{R} = \frac{1}{\omega_o RC}$	$Q = \frac{\omega_o}{B_P} = \omega_o RC = \frac{R}{\omega_o L}$

In a series RLC circuit, series resonance occurs when $X_L = X_C$. The frequency at which the resonance occurs is called resonant frequency.

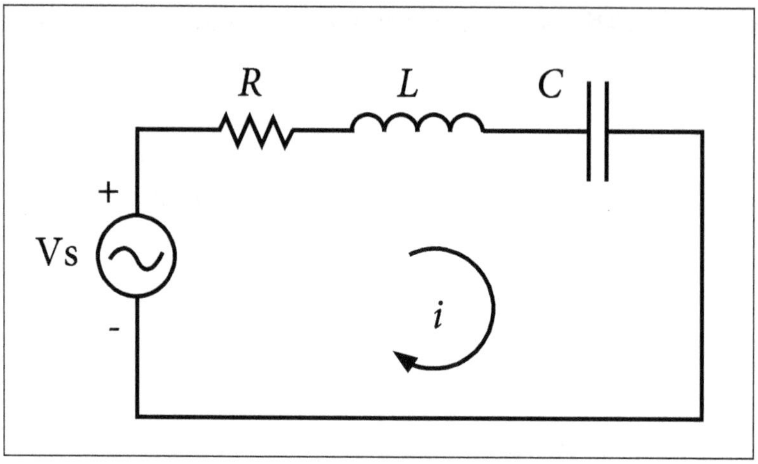

RLC Circuit.

Total impedance of the RLC circuit,

$$Z = R + jX_L - jX_C$$

$$Z = R - j\left(\frac{}{wC} - wL\right)$$

$$\text{Magnitude}|Z| = \sqrt{R^2 + \left(\frac{1}{wC} - wL\right)^2}$$

$$\text{Phase}\angle\varphi = \tan^{-1}\left(\frac{\frac{1}{wC} - wL}{R}\right) = -\tan^{-1}\left(\frac{1}{wCR} - \frac{wL}{R}\right)$$

At resonance, $\text{Magnitude}|Z| = \sqrt{R^2} = R$

Current flowing through RLC circuit $I = V/|Z|$, at resonant freq $f_r, I = V/R$

Voltage across Resistor, $V_R = I\,R$

Voltage across Inductor, $V_L = I\,X_L\,(\text{or})\; I * \omega_L$

Voltage across Capacitor, $V_C = I\,X_C\,(\text{or})\; I/\,\omega_c$

Maximum voltage occurs across the inductor above the resonant frequency,

$$f_L = \frac{1}{2\pi\sqrt{LC}}\sqrt{\frac{1}{1 - \frac{R^2C}{2L}}}$$

Maximum voltage occurs across the capacitor below the resonant frequency,

$$f_C = \frac{1}{2\pi}\sqrt{\frac{1}{LC} - \frac{R^2}{2L}}$$

Frequency Response of Resistor, Inductor and Capacitor - Series RLC Circuit

$$\text{Magnitude}|Z| = \sqrt{R^2 + \left(\frac{1}{wC} - wL\right)^2}$$

Term / Affected by frequency	R Independent of frequency		$jX_{L = j\omega L}$ Directly proportional to frequency		$-jX_{C = -j/\omega C}$ Inversely proportional to frequency	
Frequency	Magnitude $\|M\| = R$	Phase	Magnitude $\|M\| = \omega_L$	$\text{Phase} = \tan^{-1}(Im/R)$	Magnitude $\|M\| = 1/\omega C$	$\text{Phase} = \tan^{-1}(Im/R)$
$\omega = 0$	R	0°	$\|M\| = 0$	$\phi = \tan^{-1}(\omega_L/0)$ $\phi = \tan^{-1}(\infty)$ $\phi = 90^\circ$	$\|M\| = -\infty$	$\phi = -\tan^{-1}(\omega_C/0)$ $\phi = -\tan^{-1}(\infty)$ $\phi = -90^\circ$
$\omega = \infty$	R	0°	$\|M\| = \infty$		$\|M\| = -0$	

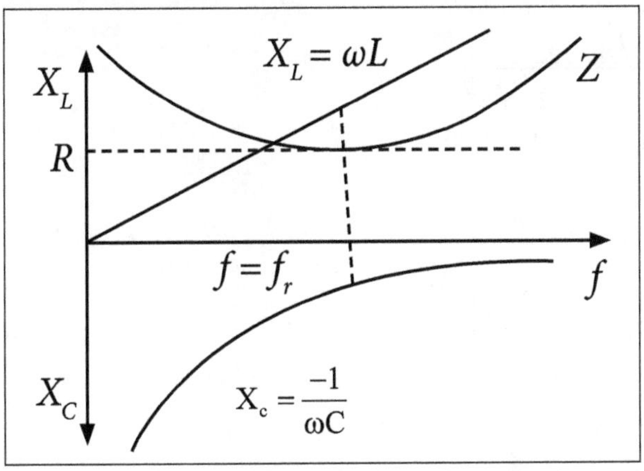

Frequency response of Impedance Z, R, XL and XC of series RLC circuit.

From the above figure:

- The reactance of capacitance X_C varies inversely with frequency.
- The reactance of X_L varies directly proportional to frequency.
- The resistor is independent of frequency variation.

3.4.2 Parallel Resonance and Bandwidth of Parallel Resonance

The basic parallel-resonant circuit is shown in the below figure. The steady state amplitude and the phase angle of the output voltage V_o vary with the frequency of the sinusoidal voltage source.

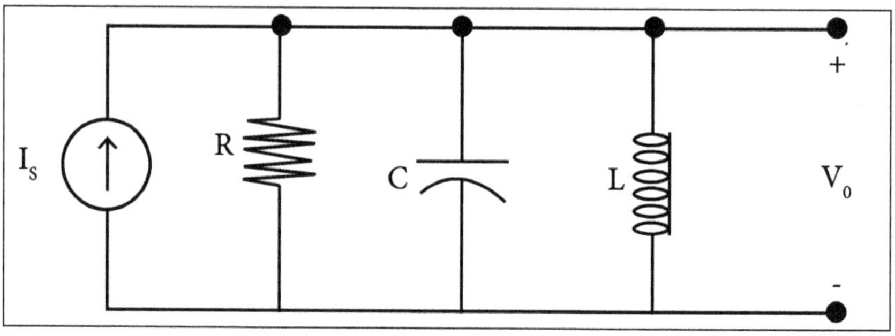

Parallel resonant circuit.

From the above figure,

$$I_s = I_m \cos(\omega t)$$

$$V_o = V_m \cos(\omega t + \theta)$$

$$\text{If } I_s = I_m \cos(\omega t), \text{ then } V_o = V_m \cos(\omega t + \theta)$$

Where,

$$V_m = \frac{I_m}{\sqrt{\frac{1}{R^2} + \left(\omega C - \frac{1}{\omega L}\right)^2}}$$

And,

$$\theta = -\tan^{-1}\left(R\left(\omega C - \frac{1}{\omega L}\right)\right)$$

The resonant frequency is,

$$\omega_o = \frac{1}{\sqrt{LC}}$$

The 3 dB frequencies are,

$$\omega_{2,1} = \sqrt{\left(\frac{1}{2RC}\right)^2 + \frac{1}{LC}} \pm \frac{1}{2RC}$$

The bandwidth,

$$B = \omega_2 - \omega_1 = 1/RC$$

The quality factor,

$$Q = \omega_o / B = R\sqrt{(C/L)}$$

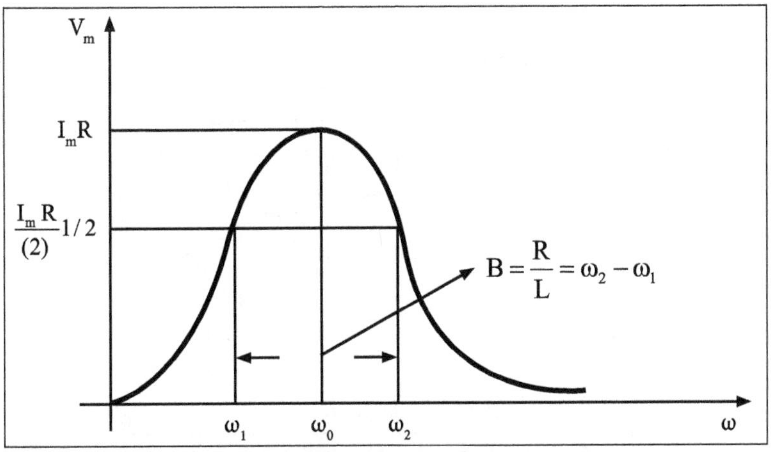

Frequency Response of Parallel Resonant Circuit.

Term / Affected by frequency	R Independent of frequency		$jX_L = j\omega_L$ Directly proportional to frequency		$-jX_C = -j/\omega_C$ Inversely proportional to frequency	
Frequency	Magnitude $\lvert M\rvert = R$	Phase	Magnitude $\lvert M\rvert = \omega_L$	$\text{Phase} = \tan^{-1}(\text{Im}/R)$	Magnitude $\lvert M\rvert = 1/\omega C$	$\text{Phase} = \tan^{-1}(\text{Im}/R)$
$\omega = 0$	R	$0°$	$\lvert M\rvert = 0$	$\phi = \tan^{-1}(\omega_L / 0)$ $\phi = \tan^{-1}(\infty)$ $\phi = 90°$	$\lvert M\rvert = -\infty$	$\phi = -\tan^{-1}(\omega_C / 0)$ $\phi = -\tan^{-1}(\infty)$ $\phi = -90°$
$\omega = \infty$	R	$0°$	$\lvert M\rvert = \infty$		$\lvert M\rvert = -0$	

If the parallel circuits impedance is at its maximum at resonance then consequently, the circuits admittance must be at its minimum and one of the characteristics of a parallel resonance circuit is that admittance is very low limiting the circuits current. Unlike the series resonance circuit, the resistor in the parallel resonance circuit has a damping effect on the circuits bandwidth making the circuit less selective.

Since the circuit current is constant for any value of impedance Z, the voltage across a parallel resonance circuit will have the same shape as the total impedance and for a parallel circuit, the voltage waveform is generally taken from the capacitor.

We know that at the resonant frequency, f_r the admittance of the circuit is at its minimum and is equal to the conductance G given by 1/R because in a parallel resonance circuit, the imaginary part of the admittance, i.e., the susceptance value, B is zero because $B_L = B_C$ as shown in the figure below.

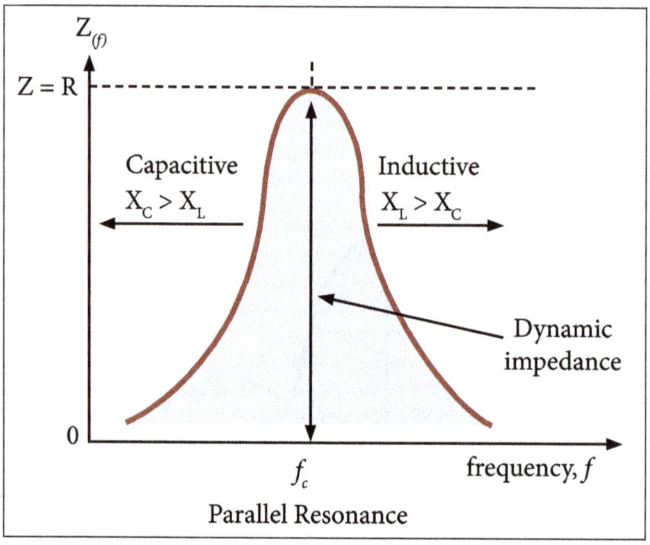

Maximum condition of parallel resonance.

Quality Factor for Parallel RLC Circuit

Circuit	Bandwidth	Capacitive	Inductive
Parallel RLC circuit - Quality factor	$Q=\frac{f_r}{f_2-f_1}=\frac{f_r}{BW}=\frac{f_r}{1/RC}=f_r RC$	$Q=wCR$	$Q=\frac{R}{wL}$

3.5 Current in Anti-resonance

As the total susceptance is zero at the resonant frequency, the admittance is at its minimum and is equal to the conductance, G. Therefore at resonance the current flowing through the circuit must also be at its minimum as the inductive and capacitive branch currents are equal ($I_L = I_C$) and are 1800 out of phase.

We remember that the total current flowing in a parallel RLC circuit is equal to the vector sum of the individual branch currents and for a given frequency is calculated as,

$$I_R = \frac{V}{R}$$

$$I_L = \frac{V}{X_L} = \frac{V}{2\pi fL}$$

$$I_C = \frac{V}{X_C} = V.2\pi fC$$

$$I_T = \text{vector sum of} \left(I_R + I_L + I_C\right)$$

$$I_T = \sqrt{I_R^2 + \left(I_L + I_C\right)}$$

At resonance, currents I_L and I_C are equal and gives a net reactive current which is equal to zero. Then, at resonance, the above equation becomes,

$$I_T = \sqrt{I_R^2 + 0^2} = I_R$$

3.6 Resistance Present in Both Branches and Anti-resonance at all Frequencies

Resistance Present in Both Branches

In some types of anti-resonant circuits, a resistance may be present in series with the capacitive branch as well as the inductive branch, as shown in figure below.

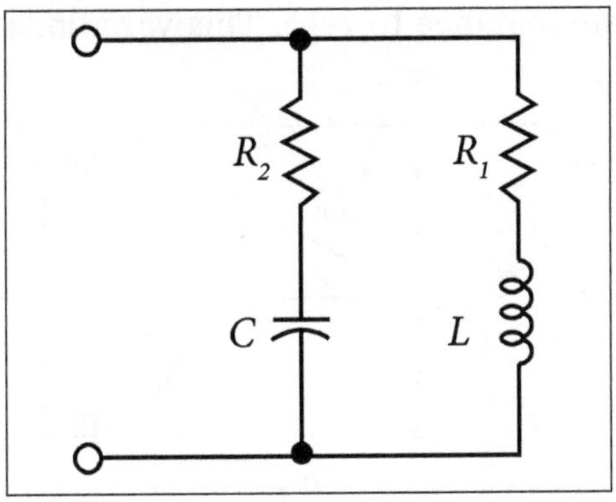

Typical form of anti-resonance circuit.

The admittance of the inductive branch is,

$$Y_L = \frac{R_1 - j\omega L}{R_1^2 + \omega^2 L^2} \qquad ...(1)$$

And that of the capacitive branch is,

$$Y_C = \frac{R_2 - j/\omega L}{R_2^2 + \frac{1}{\omega^2 C^2}} \qquad ...(2)$$

Therefore, total admittance is,

$$Y = Y_L + Y_C = \frac{R_1}{R_1^2 + \omega^2 L^2} + \frac{R_2}{R_2^2 + \frac{1}{\omega^2 C^2}} - j\left(\frac{\omega L}{R_1^2 + \omega^2 L^2} - \frac{1/\omega C}{R_2^2 + \frac{1}{\omega^2 C^2}}\right) \quad ...(3)$$

For anti-resonance, the reactive term must be zero, i.e.,

$$\omega_{ar} L\left(R_2^2 + \frac{1}{\omega_{ar}^2 C^2}\right) - \frac{1}{\omega_{ar} C}\left(R_1^2 + \omega_{ar}^2 L^2\right) = 0$$

Or,

$$f_{ar} = \frac{1}{2\pi}\sqrt{\frac{1}{LC}\left(\frac{L - R_1^2 C}{L - R_2^2 C}\right)} \quad ...(4)$$

If $R_2 = 0$, the expression reduces to, $f_{ar} = \frac{1}{2\pi}\sqrt{\frac{1}{LC} - \frac{R^2}{L^2}} = \frac{1}{2\pi}\sqrt{\frac{1}{LC}}\sqrt{1 - \frac{CR^2}{L}}$

3.6.1 Anti Resonance at all Frequencies

In case of anti-resonance, the condition for resonance was found by equating imaginary part of admittance i.e., susceptance to zero. This variations are shown in the below figure.

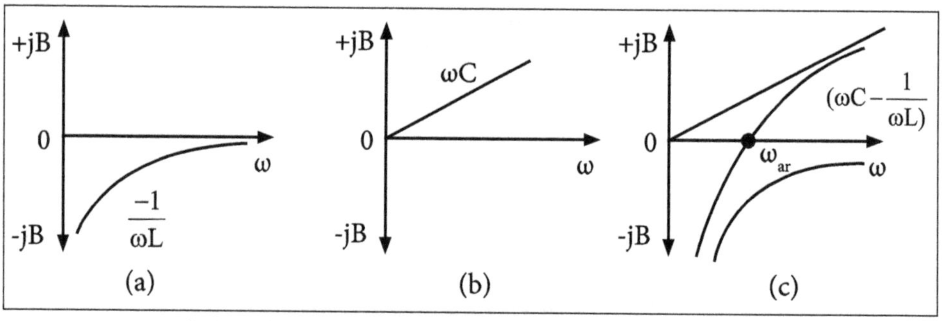

Variation of frequency on susceptance of anti-resonant circuit.

The inductive susceptance, B_L follows hyperbolic curve versus frequency as $B_L = \frac{-1}{2\pi f L}$ which is shown in the figure (a) above.

The capacitive susceptance BC follows straight line versus frequency as $B_c = 2\pi fC$ shown in the figure (b) above.

The plot of total susceptance versus frequency as shown in the above figure (c).

The point ω_{ar} on the frequency axis, where total susceptance curve crosses frequency axis is the condition of resonance. Here point of crossing of total reactance and frequency axis gives resonant frequency, ω_{ar}.

At resonant frequency far, the admittance is minimum and is equal to conductance G given by 1/R. Thus at the resonance, impedance is maximum and hence current in the circuit is minimum. The variation is as shown in figure below.

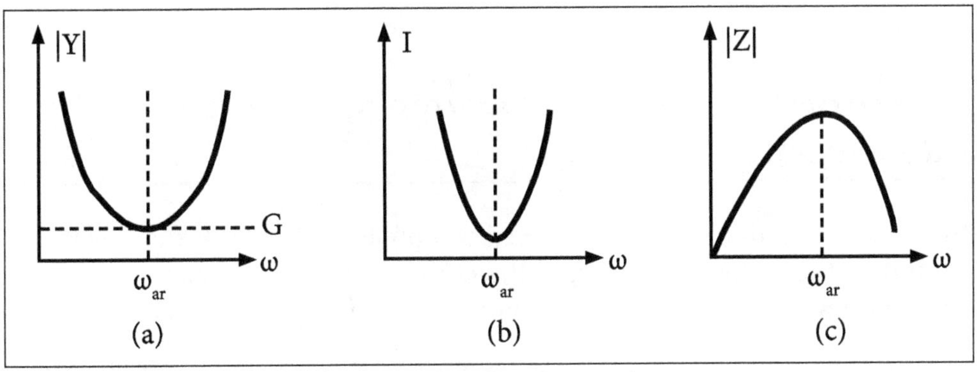

Variation of resonant frequencies.

4

Network Theorems

4.1 Network Theorems: Thevinin's, Norton's, Milliman's and Reciprocity

A network is a collection of interconnected components. Network analysis is the process of finding the voltages across and the currents through every component in the network. There are many different techniques for calculating these values.

The most important fundamental theorems of network analysis are as follows:

- Thevenin's theorem.
- Norton's theorem.
- Millman's theorem.
- Reciprocity theorem.
- Substitution Theorem.
- Superposition theorem.
- Maximum power transfer theorem.

4.1.1 Thevinin's Theorem

In circuit theory, Thevenin's theorem for linear electrical networks states that any combination of voltage sources, current sources and resistors with two terminals is electrically equivalent to the single voltage source V and the single series resistor R. For a single frequency AC systems, the theorem can also be applied to general impedances, not just for resistors.

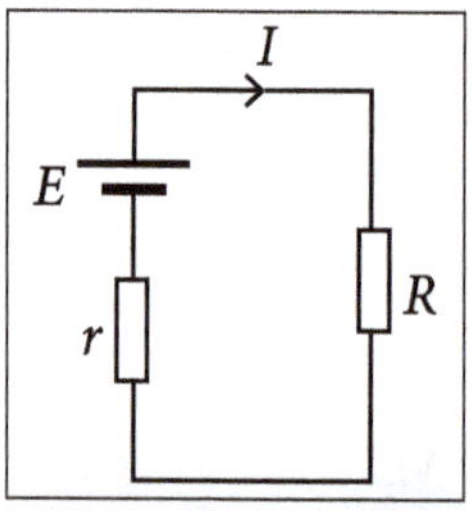

Thevenin's circuit.

The procedure adopted when using the Thevenin's theorem is summarized as follows:

- Determine the current in any branch of the active network.
- Remove the resistance R from that branch.
- Determine the open-circuit voltage E, across the break.
- Remove each of the sources of e.m.f. and replace them by their internal resistances and then determine the resistance r, 'looking-in' at the break.
- Determine the value of the current from the equivalent circuit as shown in the above figure,

  i.e., $I = ER + r$

Problem

1. Let us determine the current in the 0.8Ω resistor using Thevenin's theorem for the below circuit.

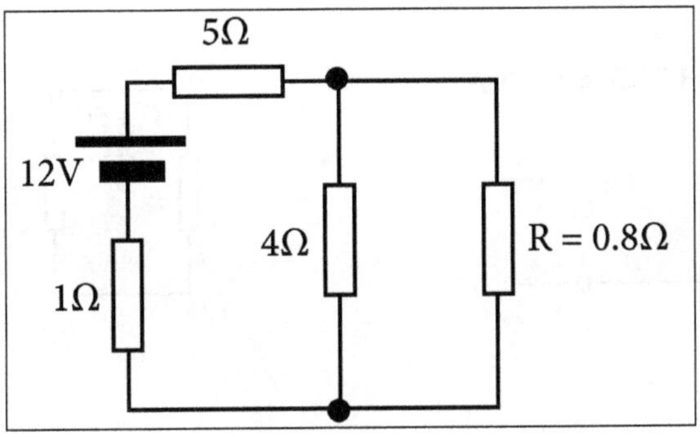

Solution:

Given:

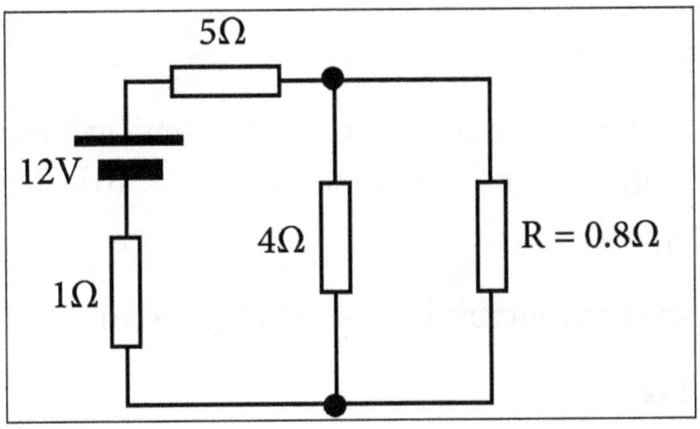

By following the above procedure,

(i) The 0.8Ω resistor is removed from the circuit as shown in the above figure.

Current, $I_1 = 12/(1+5+4)$

$= 12/10$

$= 1.2A$

(ii) P.d. across 4Ω resistor $= 4I_1 = (4)(1.2) = 4.8V$

Hence, p.d. across AB, i.e., the open-circuit voltage across AB,

$E = 4.8$ V

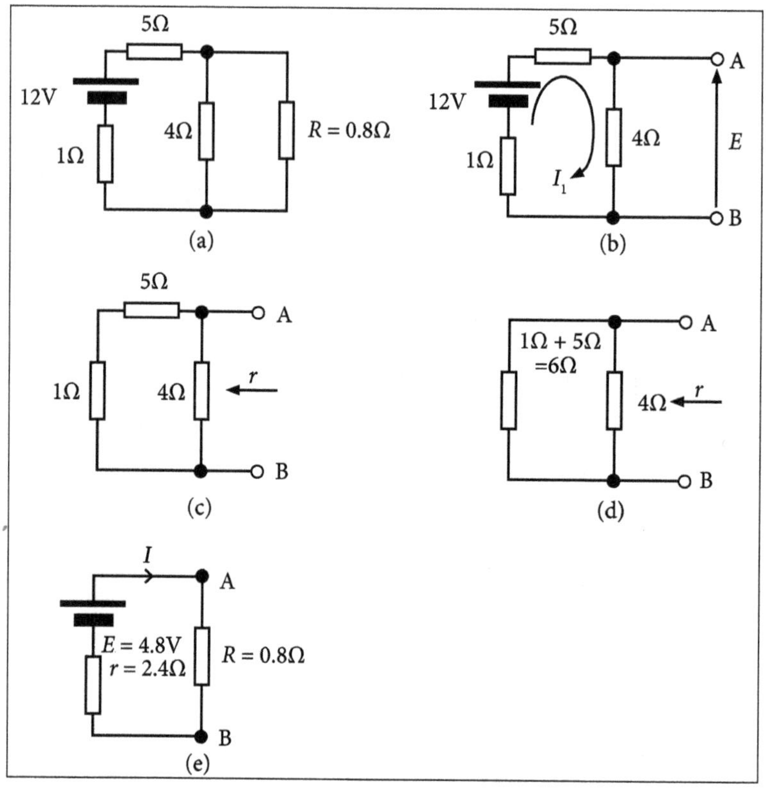

(iii) Removing the source of e.m.f. gives the circuit shown in figure (c). The equivalent circuit of figure (c) is shown in figure (d), from which resistance $r = (4 \times 6)/(4+6) = 24/10 = 2.4\ \Omega$.

(iv) The equivalent Thevenin's circuit is shown in figure (e).

From the above circuit (e),

Current $I = E/(r + R)$

$= 4.8/(2.4+0.8)$

$= 4.8 / 3.2$

$I = 1.5A =$ Current in the 0.8 Ω resistor.

4.1.2 Norton's Theorem

Norton's Theorem states that, "Any two-terminal linear bilateral DC network can be replaced by an equivalent circuit consisting of a current and a parallel resistor".

Preliminary Steps

The steps leading to the proper values of IN and RN are as follows:

- Remove the portion of the network across which the Norton equivalent circuit is found.
- Mark the terminals of the remaining two-terminal network.
- Finding R_N:
  - Let us calculate the value of R_N by first setting all sources to zero and then finding the resultant resistance between the two marked terminals. Since $R_N = R_{Th}$, the procedure and value obtained using the approach described for Thevenin's theorem will determine the proper value of R_N.
- Finding I_N:
  - Let us calculate the value of I_N by first returning all the sources to their original position and then finding the short-circuit current between the marked terminals. It is the same amount of current that can be measured by an ammeter which is placed between the marked terminals.
- Finally, draw the Norton equivalent circuit with the portion of the circuit previously removed or replaced between the terminals of the equivalent circuit.

Norton Equivalent Circuit

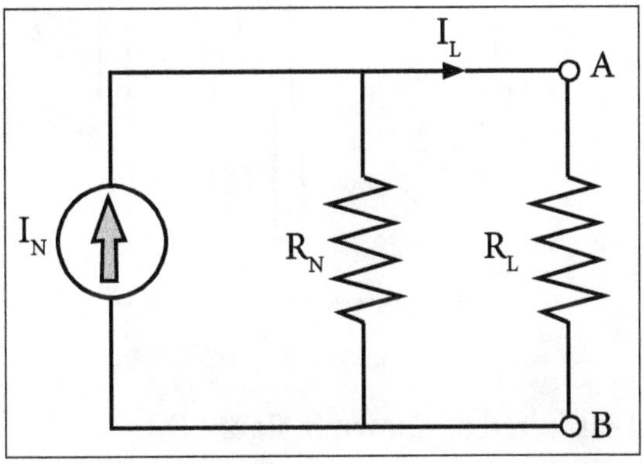

The current through R_L is given by, $I_L = \frac{I_N . R_N}{R_N + R_L}$

Where,

R_N - Norton Resistance

I_N - Norton Current

Problem

1. Let us determine the current I flowing in the 4Ω resistance shown in figure (a) by using Norton's theorem.

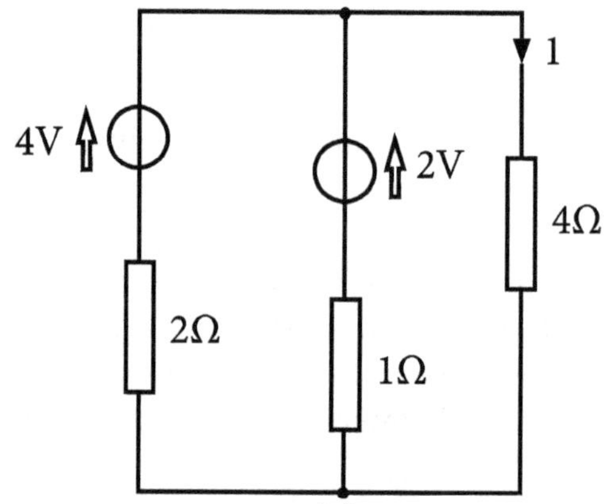

(a)

Solution:

Given:

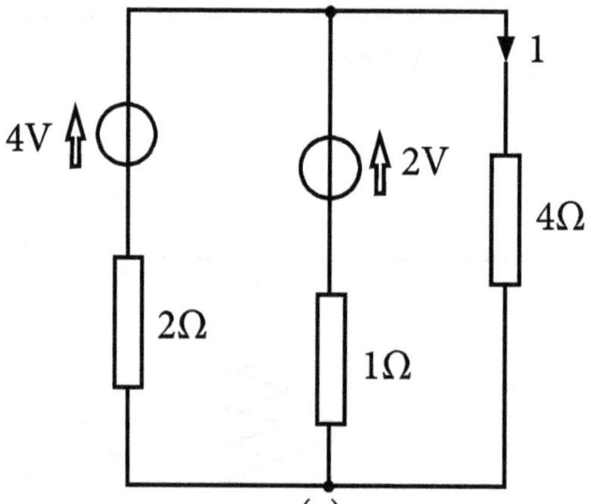

(a)

The 4Ω branch is short-circuited as shown in figure (b).

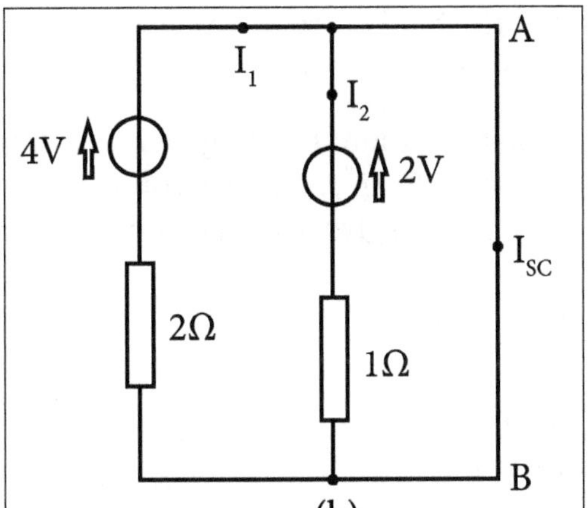

(b)

From the above figure (b),

$$I_{SC} = I_1 + I_2 = 4A.$$

If the sources of e.m.f. are removed, the resistance 'looking-in' at a break made between A and B is given by,

$$r = 2 \times 1/2 + 1 = 2/3\ \Omega$$

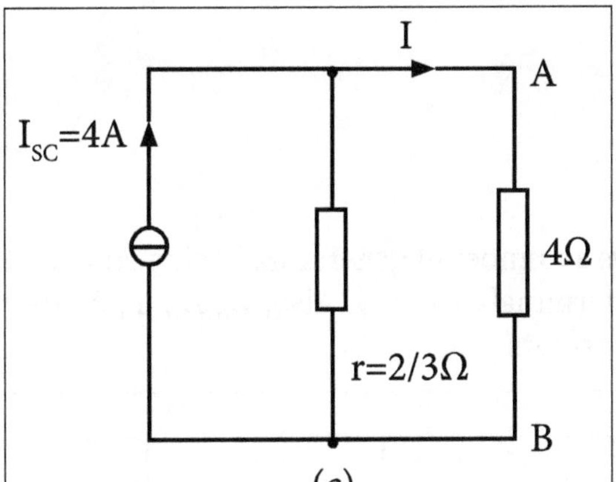

(c)

From the Norton equivalent network shown in above figure (c), the current in the 4 Ω resistances is given by,

$$I = \left[(2/3)/((2/3)+4)\right](4) = 0.571\,A.$$

4.1.3 Milliman's Theorem

It is possible to combine the number of voltage sources or current sources into a single equivalent voltage or current source using Millman's theorem. Therefore, this theorem is quite useful in calculating the total current supplied to the load in a generating station by a number of generators that are connected in parallel across a bus bar.

Millman's theorem states that "if n number of generators having generated emfs E_1, E_2, ..., E_n and internal impedances Z_1, Z_2, ..., Z_n are connected in parallel, then the emfs and impedances can be combined to give a single equivalent emf of E with an internal impedance of equivalent value Z".

Where,

$$E = \frac{E_1Y_1 + E_2Y_2 + ... + E_nY_n}{Y_1 + Y_2 + ... + Y_n}$$

$$Z = \frac{1}{Y_1 + Y_2 + ... + Y_n}$$

Where, Y_1, Y_2, ... Y_n are the admittances corresponding to the internal impedances Z_1, Z_2, ..., Z_n and are given by,

$$Y_1 = \frac{1}{Z_1}$$

$$Y_2 = \frac{1}{Z_2}$$

$$\vdots$$

$$Y_n = \frac{1}{Z_n}$$

The below figure shows a number of generators having emfs E_1, E_2, ..., E_n are connected in parallel across the terminals x and y. Also, Z_1, Z_2, ..., Z_n are the respective internal impedances of the generators.

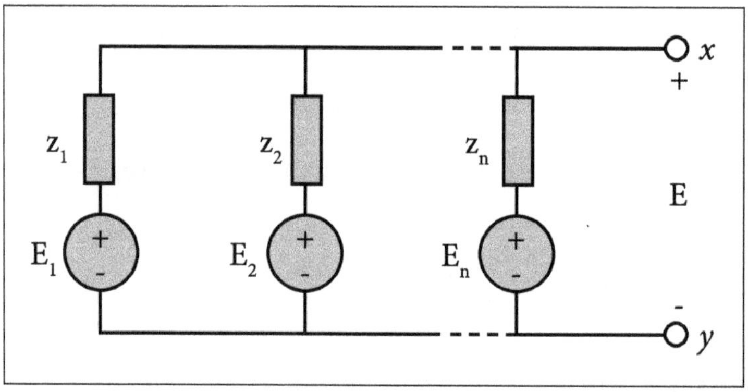

The Thevenin equivalent circuit of the above figure using Millman's theorem is shown in the below figure.

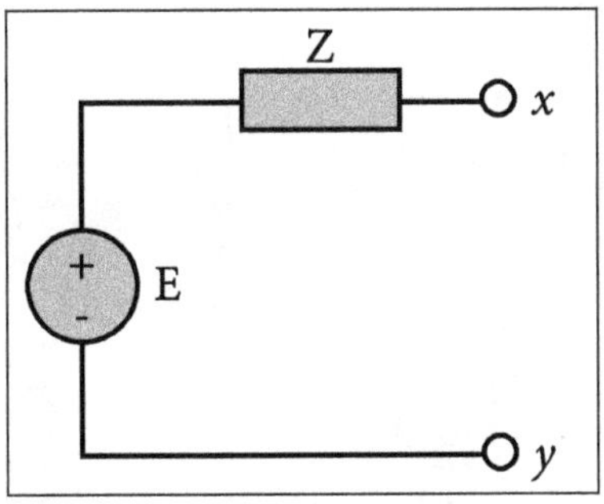

Thevenin equivalent circuit.

The nodal equation at x gives,

$$\frac{E_1 - E}{Z_1} + \frac{E_2 - E}{Z_2} + ... + \frac{E_n - E}{Z_n} = 0$$

$$\Rightarrow \quad \left[\frac{E_1}{Z_1} + \frac{E_2}{Z_2} + ... + \frac{E_n}{Z_n}\right] = E\left[\frac{1}{Z_1} + \frac{1}{Z_2} + ... + \frac{1}{Z_n}\right]$$

$$\Rightarrow \quad E_1 Y_1 + E_2 Y_2 + ... + E_n Y_n = E\left[\frac{1}{Z}\right]$$

Where,

Z = Equivalent internal impedance.

or,

$$[E_1 Y_1 + E_2 Y_2 + ... + E_n Y_n] = EY$$

$$\Rightarrow \quad E = \frac{E_1 Y_1 + E_2 Y_2 + ... + E_n Y_n}{Y}$$

Where,

$$Y = Y_1 + Y_2 + ... + Y_n$$

And,

$$Z = \frac{1}{Y} = \frac{1}{Y_1 + Y_2 + ... + Y_n}$$

Problem

1. Referring to the circuit shown in the below figure, let us determine the current through 10 Ω resistor using Millman's theorem.

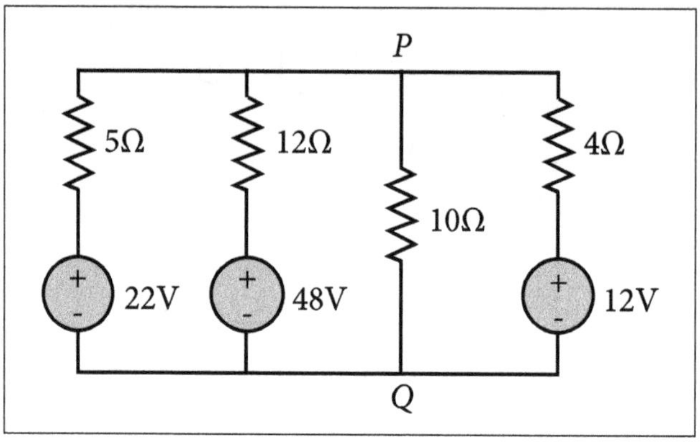

Solution:

Given:

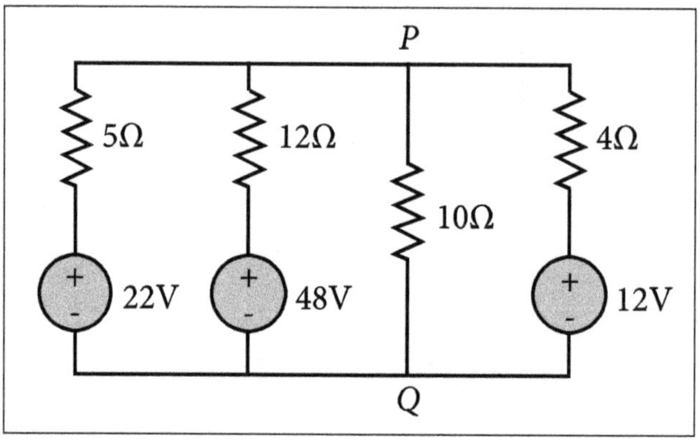

Using Millman's theorem, the circuit shown in the given figure is replaced by its Thevenin equivalent circuit across the terminals PQ as shown in the below figure.

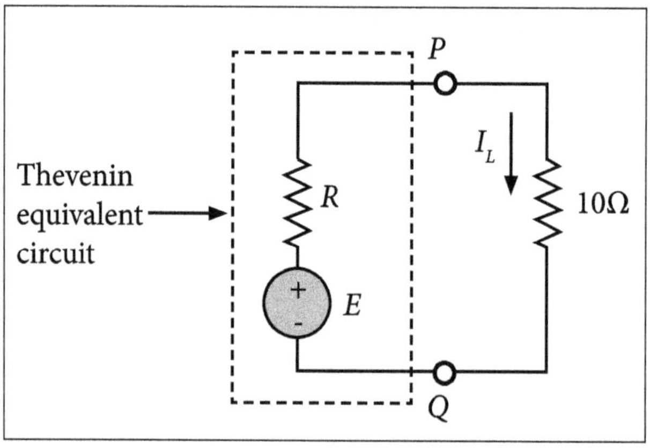

$$E = \frac{E_1 Y_1 + E_2 Y_2 - E_3 Y_3}{Y_1 + Y_2 + Y_3}$$

$$= \frac{22\left(\frac{1}{5}\right)+48\left(\frac{1}{12}\right)-12\left(\frac{1}{4}\right)}{\frac{1}{5}+\frac{1}{12}+\frac{1}{4}}$$

$$=10.13 \text{ Volts}$$

$$R=\frac{1}{Y_1+Y_2+Y_3}$$

$$=\frac{1}{0.2+0.083+0.25}$$

$$=1.88\ \Omega$$

Hence,

$$I_L=\frac{E}{R+10}=0.853\,A$$

4.1.4 Reciprocity Theorem

Reciprocity theorem is one of the DC network analysis and AC network analysis technique and deals with the relationship between impressed source in a part of the circuit and its response at some other part of the circuit.

Reciprocity theorem states that, in any branch of a network or circuit, the current due to a single source of voltage (V) in the network is equal to the current through that branch in which the source was originally placed when the source is again put in the branch in which the current was originally obtained.

The supply voltage "V" and a 0-ohm ammeter reading or current "I" are mutually transferable in any bilateral linear circuits. The ratio between the Voltage and the Current that are mutually transferable is called the transfer resistance.

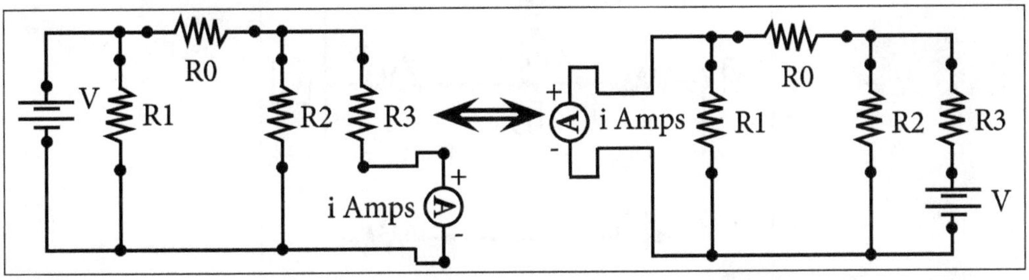

The Reciprocity theorem is true in case of bilateral linear circuits or on the linear circuits which contains only bilateral components. This property of bilateral linear circuits is termed as their reciprocal property and the circuits in which the Reciprocity theorem holds true are termed as reciprocal circuits.

Reciprocity theorem can be applied to solve many DC and AC electrical network efficiently and it also have special applications in the field of electromagnetism and antenna electronics.

Problem

1. The current in the various branches of the network shown in figure below. Let us determine the current through the one volt battery (V) when an extra e.m.f of 1 volt is added on the branch BD opposing the flow of original current in that branch.

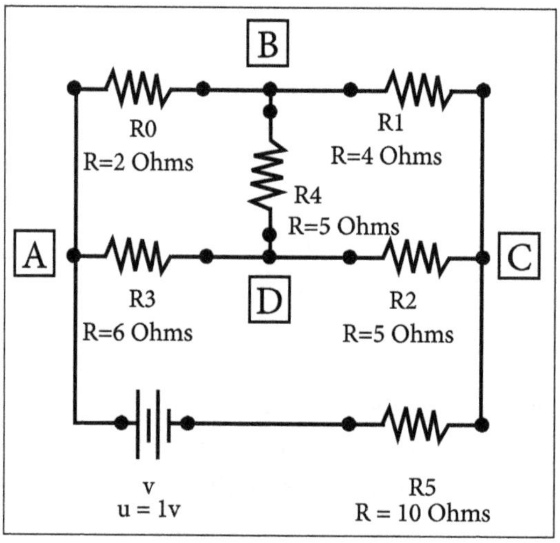

Solution:

Given:

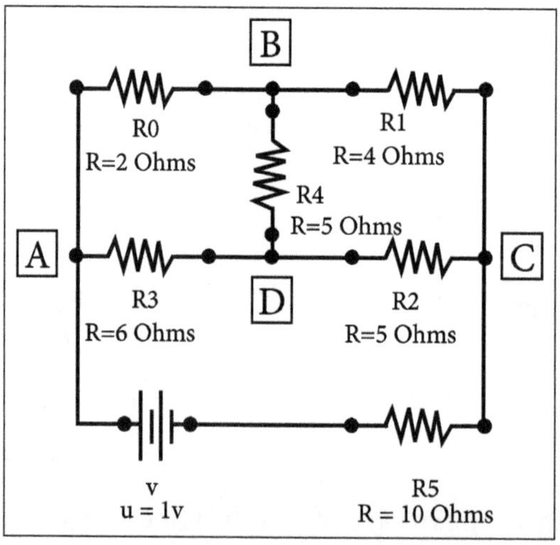

First solve the circuit using the Mesh Analysis or Loop Current Method,

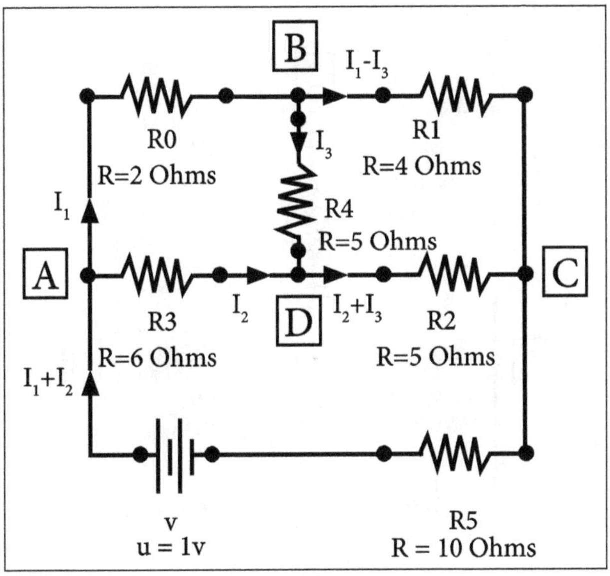

On Loop ABDA,

$$-2I_1 - 8I_3 + 6I_2 = 0$$

or,

$$I_1 - 3I_2 + 4I_3 = 0$$

On Loop BCDB,

$$-4(I_1 - I_3) + 5(I_2 + I_3) + 8I_3 = 0$$

or,

$$4I_1 - 5I_2 - 17I_3 = 0$$

On Loop ABCVA,

$$-2I_1 - 4(I_1 - I_3) - 10(I_1 + I_2) + 1 = 0$$

or,

$$16I_1 + 10I_2 - 4I_3 = 1$$

Thus, solving these three equations, we get,

$$I_1 = 0.0494A,\ I_2 = 0.0229A \text{ and } I_3 = 0.0049A$$

Now, as given in question let us add an extra e.m.f of 1 volt on branch BD opposing the flow of original current in that branch,

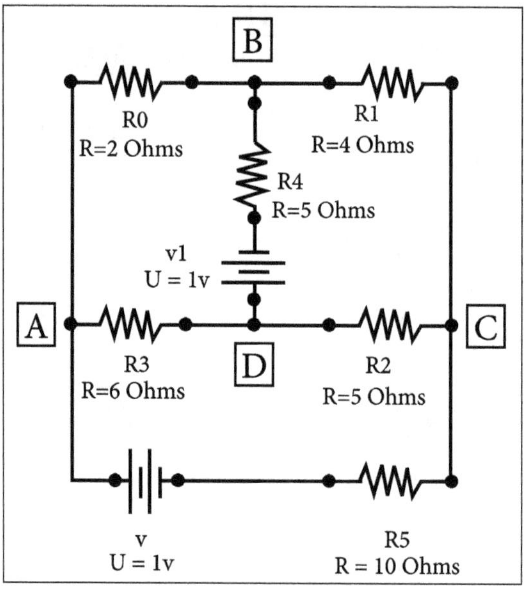

Now using the Superposition Theorem, let us find the current through the battery V and V_1 separately.

Due to V

Current through V due to $V = I_1 + I_3 = 0.0723$ A

Due to V_1

Now let us again use reciprocity theorem. The one voltage supply at V produced I_3 at branch BD, so when -1 voltage is acted on the branch BD the current induced in the place of V will be $-I_3$.

So current through V due to $V_1 = -I_3 = -0.0049$ A

Thus, the current thorough V due to both voltages is, 0.0723 A − 0.0049 A = 0.0674 A

4.2 Compensation and Substitution Theorem

In any bilateral active network, if the amount of impedance is changed from the original value to some other value carrying a current of I, then the resulting changes that occurs in other branches are same as those that would have been caused by the injection voltage source in the modified branch with a negative sign, i.e., minus of the voltage current and it changed the impedance product.

4.2.1 Substitution

Substitution theorem states that any branch in a network can be substituted by a different branch without disturbing the currents and voltages in the whole network provided that the new branch has the same set of terminal voltages and current as the original branch. Substitution theorem can be used in both the linear and nonlinear circuits.

4.3 Superposition Theorem

Superposition Theorem is one of the electrical network analysis theorems which helps us to solve the linear circuit with more than one current or voltage source easily.

The superposition theorem for electrical circuits states that for a linear system the response (voltage or current) in any branch of a bilateral linear circuit having more than one independent source equals the algebraic sum of the responses caused by each independent source acting alone, where all the other independent sources are replaced by their internal impedances.

Suppose an electrical circuit having several branches or loads and also several source some being current source and some being voltage source.

Then Superposition theorem suggests that, if we find the branch responses on a branch due to only of those source by ignoring the effect of all the other sources or replacing all the other sources by their corresponding internal impedance and repeat the process for every source on the circuit, then the combined responses (Voltage drop and Current through it) on a branch due to all the sources combined is the algebraic sum of the responses on the branches due to each individual sources.

Steps to be Followed to Apply the Superposition Theorem

Step 1:

Select any one of the energy source and replace all the other energy sources by their internal series resistances.

Step 2:

With only one energy source, let us calculate the current through the required branch using series parallel along with current division rule or else using mesh analysis.

Step 3:

Repeat the steps 1 and 2 for each source individually.

Step 4:

Algebraically add the currents obtained due to the individual source to obtain the combined effect of all the sources.

Problem

1. Let us determine the current I_{ab} flowing through the 3Ω resistor as shown in the below figure using the superposition theorem.

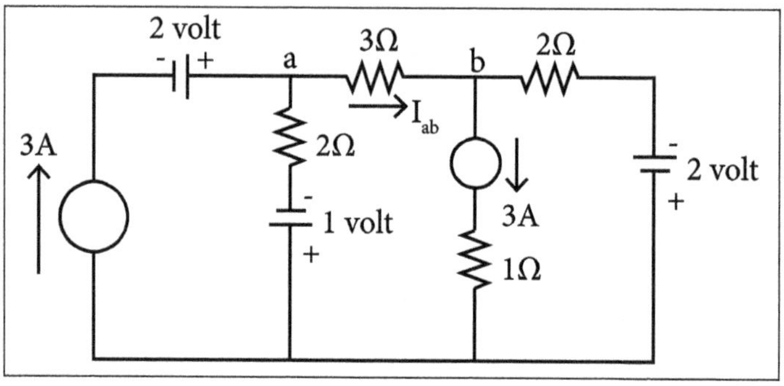

Solution:

Given:

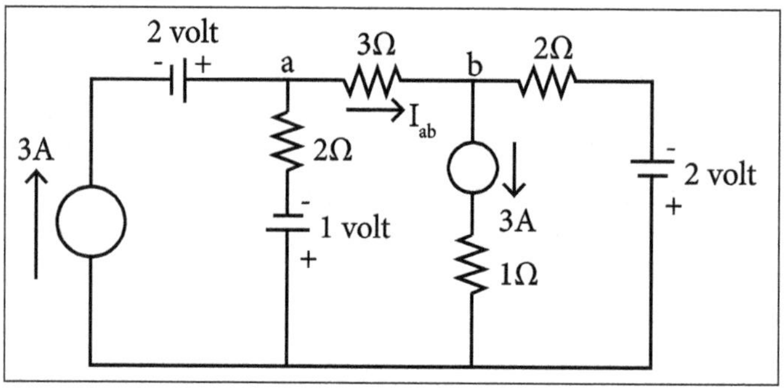

(a)

Let us assume that the current source 3A is acting alone in the circuit and the internal resistances replace the other sources.

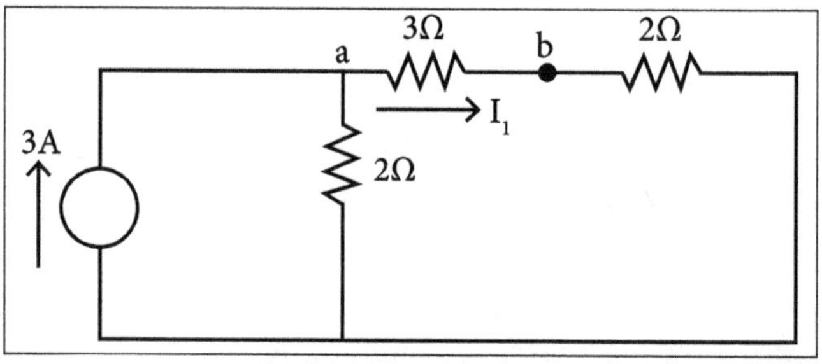

(b)

The current flowing through 3Ω resistor can be obtained from figure (b) and it is given by,

$$I_{1(\text{due to 3 A current source})} = 3\times\frac{2}{7} = \frac{6}{7}\,A\,(a\text{ to }b) \qquad ...(1)$$

Current flowing through 3Ω resistor due to 2V source can be obtained from figure (c).

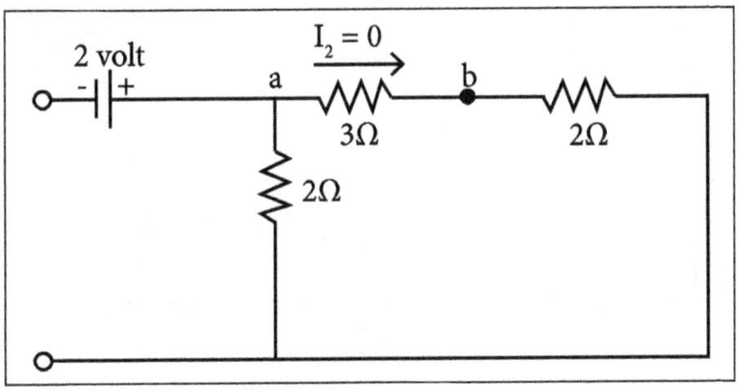

(c)

It is seen that no current is flowing.

$$I_{2(\text{due to 2V voltage source})} = 0A\ (a\text{ to }b)$$

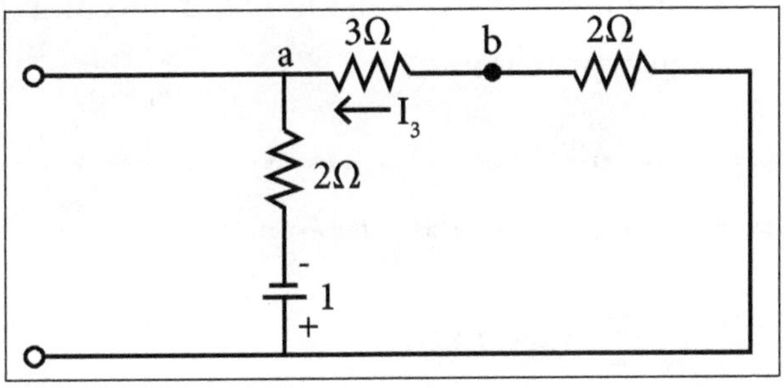

(d)

Current through 3Ω resistor due to 1V voltage source only is given by,

$$I_{3(\text{due to 1V voltage source})} = 1/7\,A\,(b\text{ to }a)$$

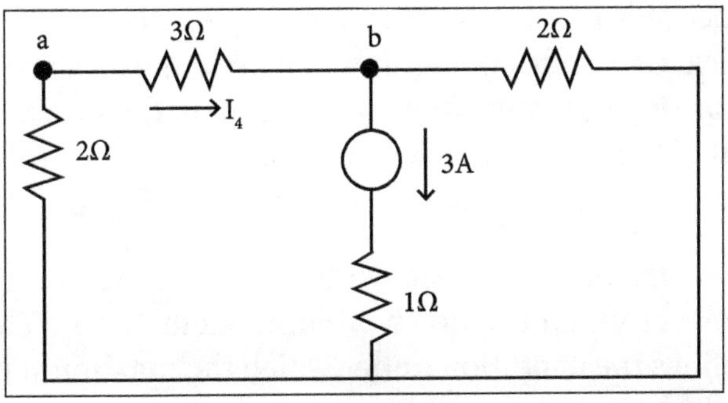

(e)

From above figure (e), Current through 3Ω resistor due to 3A current source only is obtained by,

$$I_{4(\text{due to 3 A current source})} = 3 \times \frac{2}{7} = \frac{6}{7} A(a \, to \, b)$$

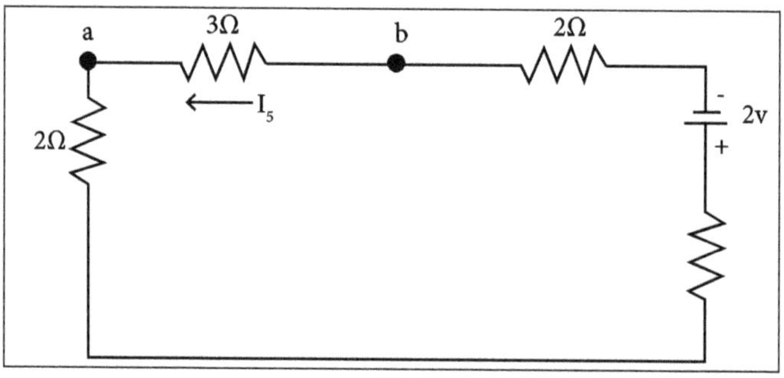

(f)

From figure (f), Current through 3 Ω resistor due to 2V voltage source only is given by,

$$I_{5(\text{due to 2V voltage source})} = \frac{2}{7} A \text{ (b to o)} \qquad ...(2)$$

Resultant current I_{ab} flowing through 3Ω resistor due to the combination of all sources is obtained by the following expression,

$$I_{ab} = I_{1(\text{due to 3A current source})} + I_{2(\text{due to 2V current source})} + I_{3(\text{due to 1V current source})}$$
$$+ I_{4(\text{due to 3A current source})} + I_{5(\text{due to 3A current source})}$$

$$= \frac{6}{7} + 0 - \frac{1}{7} + \frac{6}{7} - \frac{2}{7} = \frac{9}{7} = 1.285(a \text{ to } b)$$

4.4 Max Power Transfer Theorems

Maximum power theorem is a theorem or technique used in the Electrical Network Analysis and Electrical circuit designing. It was invented by a German engineer Moritz von Jacobi in the year 1840. He invented the theorem in the process of finding a way to maximize the output of the battery to a motored boat which he designed to travel in the river Neva. Hence, the theorem is also sometimes referred as Jacobi's Law.

Maximum power transfer theorem deals with the power transferred to the load on a circuit with a network of various sources or components on it. The maximum power transfer theorem defines the condition under which the maximum power is transferred to the load in a circuit.

Under the condition of Maximum power transfer, we only deal with the power transferred to the load and does not consider the power dissipated in internal circuits or resistance of the source. Thus, we are not considering about the maximum efficiency of power transfer, but instead maximum possible power transfer from a source to a load.

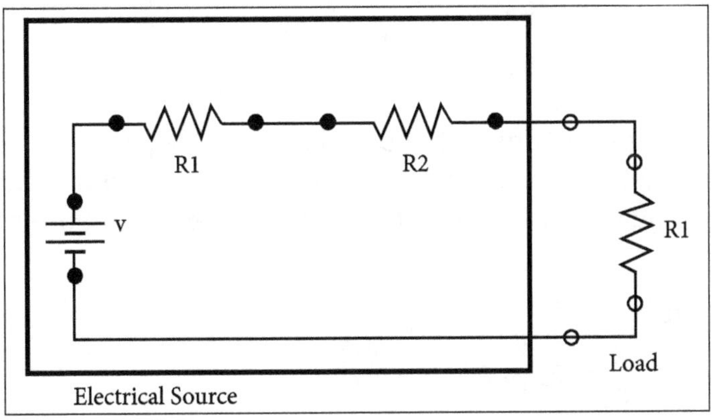

Maximum power transfer theorem – example.

According to maximum power transfer theorem, maximum power is yielded to the load RL when RL is equal to the internal resistance of the circuit or R1+R2. The Maximum power transfer theorem is true in case of any kind of circuit if it may be linear, non-linear, active, DC or AC.

In case of DC circuits, load resistance is matched with the internal resistance of the source by making both resistance equal and in case of AC, the load impedance is matched with the internal impedance of the circuit or source by making the load impedance as the complex conjugate of the source impedance.

For example, load impedance will be R1 - jX if the internal impedance of the source is R1+ jX.

Proof of Maximum Power Transfer Theorem

Let,

V = EMF supplied to the load.

R_L = Load resistance.

R_i = Internal resistance of the source.

I = Current flowing through the load, internal resistance and the source of the circuit.

P_L = Power transferred to the load.

P_i = Power dissipated at the internal resistance.

Then,

Power transferred to the load $= P_L = I^2 R_L$

$$P_L = \left(\frac{V}{R_i + R_L}\right)^2 \times R_L = \frac{V^2}{\frac{R_i^2}{R_L} + 2R_i + R_L}$$

Now using the theorems of Differential calculus, if we keep the value of R_L as variable and want to calculate the maximum value of P_L, then we need to differentiate P_L with respect to R_L and equate it with zero. Thus, under Maximum power transfer to load condition,

$$\frac{d}{dR_L} P_L = \frac{d}{dR_l} \frac{V^2}{\frac{R_i^2}{R_L} + R2_i + R_L} = 0$$

Or,

$$\frac{R_i^2}{R_L^2} + 1 = 0$$

Or,

$$R_i = R_L$$

And in AC networks using same mathematical technique, we can prove.

If,

$Z_i = R_i + X_i =$ Internal impedance of reactive circuits.

$Z_L = R_L + X_L =$ Load impedance.

Then, under the condition of maximum power transfer to load,

$$Z_i = Z_L,\ R_i = R_L \text{ and } X_i = -X_L$$

Power Transfer Efficiency

Power transfer efficiency is the efficiency of any source or circuit in transferring its power to the load or it is the ratio of power transferred to the load over total power transferred by the source. It is denoted by the Greek letter η.

Mathematically,

$$\eta = P_L / P_T$$

Where,

P_L is the power transferred to the load.

P_T is the total power transferred by the source.

We can expand this expression as,

$$P_L = I^2R_L$$

$$P_T = P_L + P_I = I^2R_L + I2R_I$$

Where,

I is the total current flowing through the circuit or Thevenin's equivalent of the circuit.

P_I is the power dissipated in internal circuits of the source.

Thus,

$$\eta = \frac{P_L}{P_T} = \frac{I^2R_L}{I^2R_L + I^2R_I} = \frac{R_L}{R_L + R_I} = \frac{1}{1 + \frac{R_I}{R_L}}$$

As we know that under the condition of maximum power transfer $R_L = R_I$, we can derive that the efficiency under the condition of maximum power transfer is only 0.5.

The overall efficiency decreases if the value of R_L is kept very low and it increases up to one when the value of R_L is increased to infinity as shown on the graph below and the power transferred to load becomes minimum when R_L is kept very low and very high.

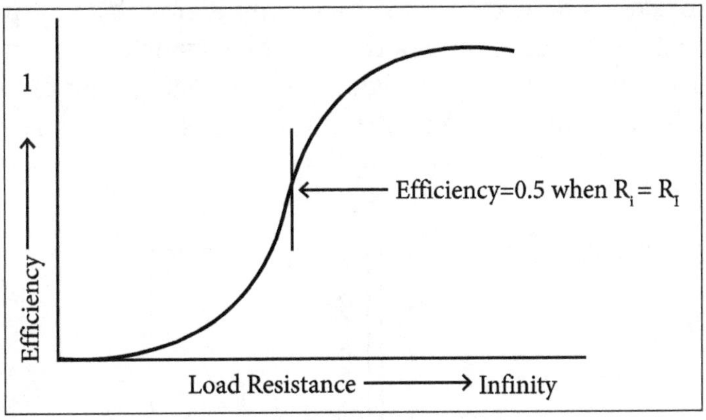

Power transfer efficiency Vs Load resistance.

The maximum power transfer to load is obtained when $R_L = R_I$ as shown in the following graphs.

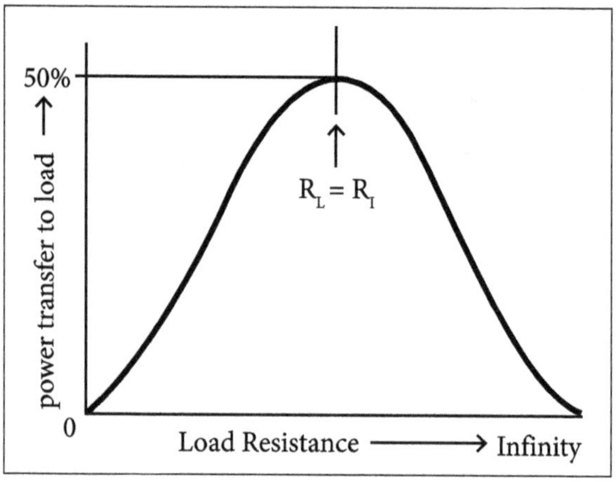

Power transfer to load Vs Load Resistance.

Applications of Maximum Power Transfer Theorem

The Maximum Power Transfer Theorem has a wide range of usage on real life situation. This theorem is used to maximize the power output to a load from any circuit. So, these can be used to design circuits where the maximum output performance is desired for example to match the amplifier with the loudspeaker to yield the maximum power to the speaker and therefore, it produces maximum sound.

In some other situations, Transformer Coupling are also used to yield maximum power to the load when the matching of Load and Source impedance is not possible. The application of Maximum Power Theorem is done only under the conditions when the maximum performance is desired over the overall efficiency of the circuit because the efficiency of the circuit under maximum power transfer condition is only 0.5.

Problems

1. If an 8Ω loudspeaker is to be connected to an amplifier with an output impedance of 1000Ω, let us calculate the turns ratio of the matching transformer required to provide maximum power transfer of the audio signal. Let us assume that the amplifier source impedance is Z_1, the load impedance is Z2 with the turns ratio given as N.

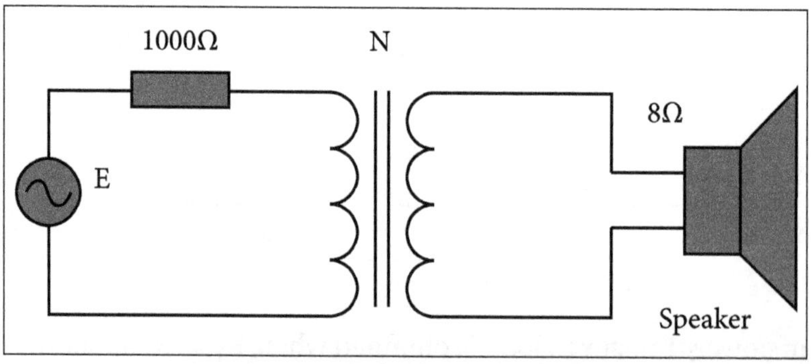

Transformer impedance matching circuit.

Solution:

Given:

8Ω loudspeaker

Output impedance = 1000Ω

Amplifier source impedance = Z_1

Load impedance = Z_2

Transformer turns ratio,

$$Z_1 = N^2 Z_2 \quad \therefore \quad N = \sqrt{\frac{Z_1}{Z_2}}$$

Therefore,

$$N = \sqrt{\frac{Z_1}{Z_2}} = \sqrt{\frac{1000}{8}} = 11.2:1$$

Generally, small transformers used in low power audio amplifiers are usually regarded as ideal. So, any losses can be ignored.

2. Let us determine the value of RL for the given network below that the power is maximum. And also determine the Max Power through load-resistance R_L by using maximum power transfer theorem.

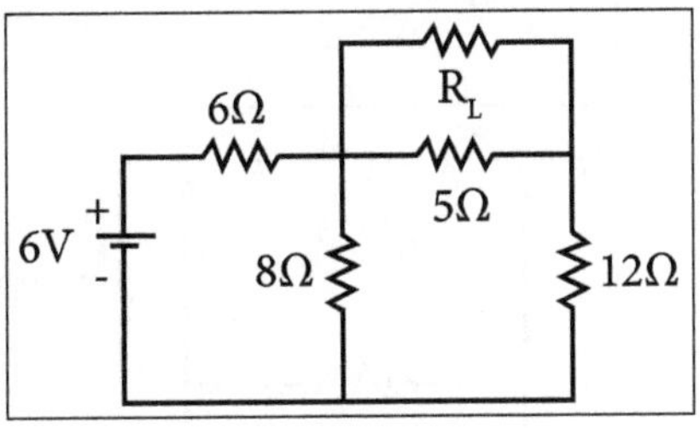

Solution:

Given:

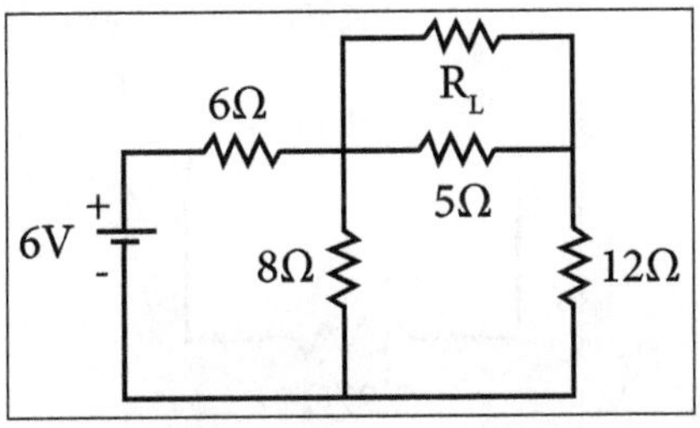

For the above network, we are going to determine the value of unknown resistance called "R_L". When power is maximum through load, resistance is equal to the equivalent resistance between the two ends of load-resistance after removing. So, for finding load resistance R_L, we have to determine the equivalent resistance for this circuit.

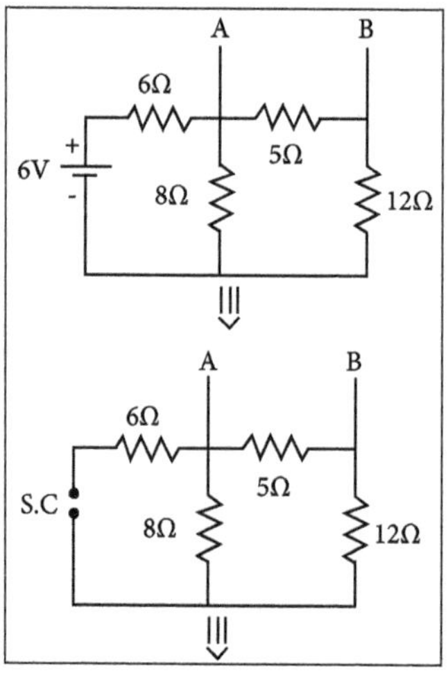

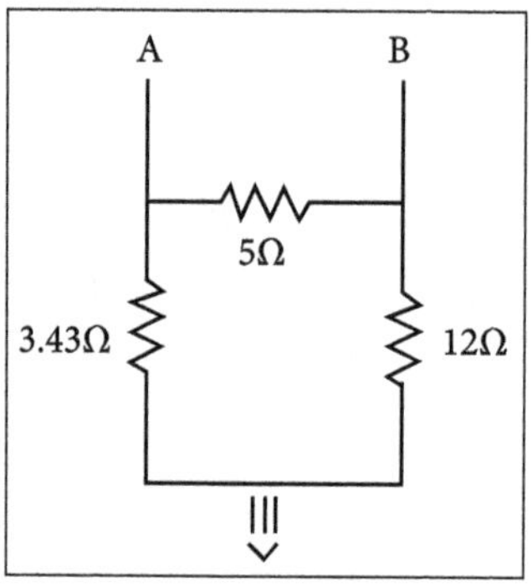

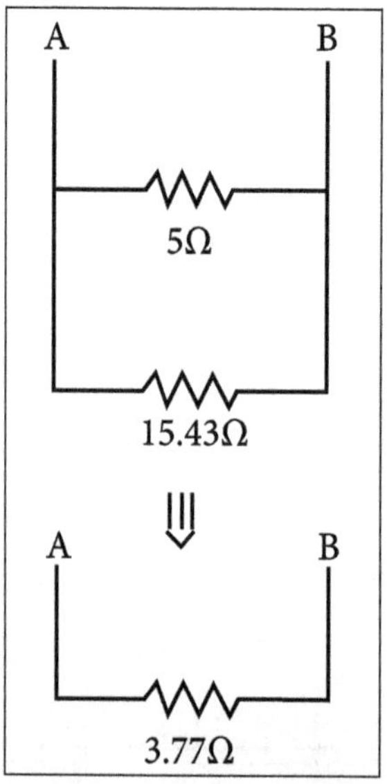

Therefore,

$$R_{AB} / R_L = 3.77\Omega$$

Now, for finding Maximum Power through load-resistance, we have to find-out the value of $V_{o.c}$.

Here, $V_{o.c}$ is known as voltage between open circuits.

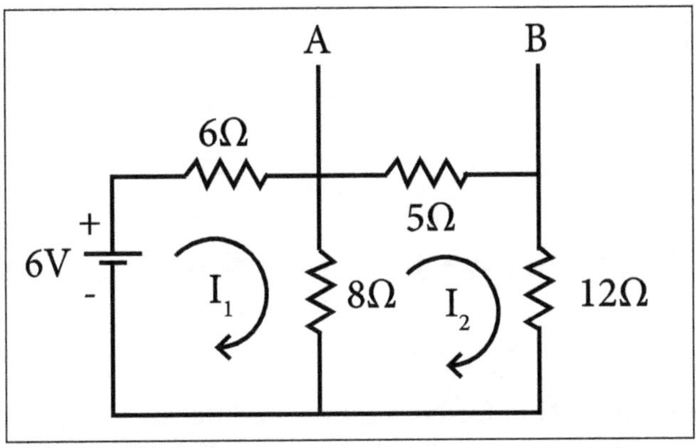

Applying KVL in 1st loop, we get as,

$$6 - 6I_1 - 8I_1 + 8I_2 = 0$$

$$-14I_1 + 8I_2 = -6 \qquad ...(1)$$

Again, Applying KVL in 2nd loop, we get

$$-8I_2 - 5I_2 - 12I_2 + 8I_1 = 0$$

$$8I_1 - 25I_2 = 0 \qquad ...(2)$$

On solving equations (1) and (2), we get as,

$$I_1 = 0.524 \text{ A}$$

$$I_2 = 0.167 \text{ A}$$

Now, from the circuit $V_{o.c}$ is,

$$V_A - 5I_2 - V_B = 0$$

$$V_{o.c} / V_{AB} = 5I_2$$

$$= 5 \times 0.167$$

$$= 0.835V$$

So, the maximum power through R_L is given by,

$$P_{max} = \frac{V_{o.c}^2}{4R_L}$$

$$P_{max} = \frac{0.835^2}{4 \times 3.77}$$

$$P_{max} = 0.046 \text{ watt}$$

4.5 Tellegen's Theorem: Problem Solving using Dependent Sources

Tellegen's theorem is applicable for circuits with a liner or nonlinear, passive or active and hysteric or non-hysteric networks. It states that the summation of the instantaneous power in circuit with n number of branches is zero.

Tellegen's theorem is applicable to a wide range of electrical networks. The only requirement for the validation of the Tellegen's theorem in any circuit is that it satisfies the Kirchhoff's Current Law and Kirchhoff's Voltage Law.

Mathematically the Tellegen's Theorem states that,

$$\sum_{k=1}^{n} P_k = V_k \times I_k = 0$$

i.e.,

$$\sum_{k=1}^{n} v_k \cdot i_k = 0$$

Where,

P_k is the instantaneous power in k'th element of the circuit.

n is the numbers of elements in the circuit.

I_k is the instantaneous current flowing through k'th element of the circuit.

V_k is the instantaneous voltage drop in k'th element of the circuit.

4.5.1 Problem Solving using Dependent Sources

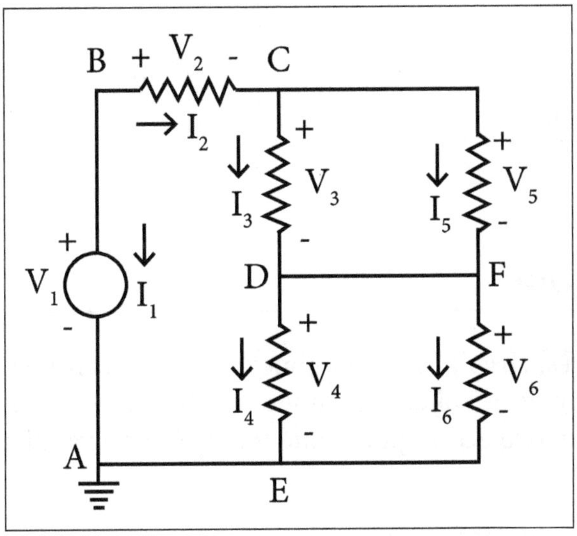

Let us consider that V_1, V_2 and V_3 be 7, 2 and 3 volts respectively.

Applying Kirchhoff voltage law around loop ABCDEA, we can found that V_4 = 2 volt is required. Around loop CDFC, V_5 is required to be 3 volt and around loop DFED, V6 is required to be 2. Next let us apply Kirchhoff current law successively to nodes B, C and D.

At node B, let i_1 = 5 A, then it is required that i_2 = – 5 A.

At node C, let i_3 = 3 A and then i_5 is required to be – 8.

At node D, assume that i_4 to be 4 then i_6 is required to be – 9.

Carrying out the operation of equation,

$$\sum_{k=1}^{n} v_k \cdot i_k = 0$$

We get,

$$7\times5+2\times(-5)+3\times3+2\times4+3\times(-8)+2\times(-9)=0.$$

Hence, Tellegen theorem is verified.

5

Two Port Networks

5.1 Introduction

RF/microwave devices, circuits and components can be classified as one, two, three or N-port networks. A majority of circuits under analysis are two-port networks. A two-port network has only two access ports, one for input or excitation and one for output or response.

A microwave circuit is formed, when several microwave devices and components such as sources, attenuators, resonators, filters, amplifiers, etc. are coupled together by transmission lines or wave guides for the desired transmission of microwave signal.

5.1.1 Relationship of Two Port Networks, Z-parameters, Y-parameters, Transmission Line Parameters, H-parameters

Low-Frequency Parameters

To characterize a linear network at low frequencies, several different sets of parameters are available. Low frequency circuits can be described by two port networks and their parameters such as Z, Y, H, ABCD, etc., as per network theory. Here, network parameters relate the total voltages and total currents as shown in below figure.

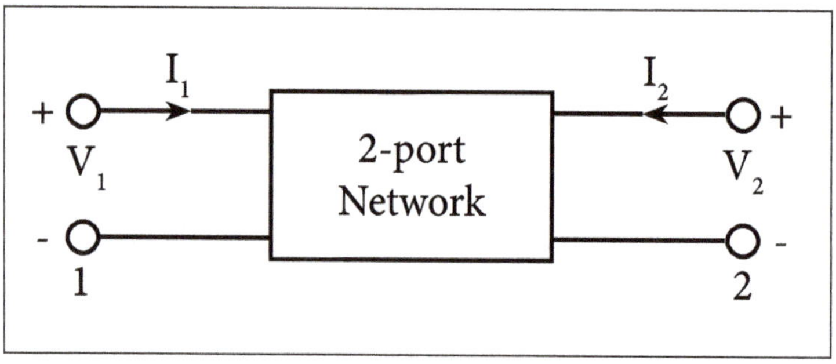

Block diagram of a two port network.

1. Impedance or Z-parameters

$$V_1 = Z_{11}I_1 + Z_{12}I_2 \qquad ...(1)$$

$$V_1 = Z_{22}I_1 + Z_{22}I_2 \qquad ...(2)$$

In matrix form,

$$[V]=[Z][I] \quad ...(3)$$

$$[V]=\begin{bmatrix} V_1 \\ V_2 \end{bmatrix}$$

$$[I]=\begin{bmatrix} I_1 \\ I_2 \end{bmatrix}$$

$$\begin{bmatrix} V_1 \\ V_2 \end{bmatrix}=\begin{bmatrix} Z_{11} & Z_{12} \\ Z_{21} & Z_{22} \end{bmatrix}\begin{bmatrix} I_1 \\ I_2 \end{bmatrix} \quad ...(4)$$

$$z_{11}=\left.\frac{V_1}{I_1}\right|_{I_2=0}, \quad z_{12}=\left.\frac{V_1}{I_2}\right|_{I_1=0}, \quad z_{21}=\left.\frac{V_2}{I_1}\right|_{I_2=0}, \quad z_{22}=\left.\frac{V_2}{I_2}\right|_{I_1=0}$$

In the preceding equations, letting I_1 or $I_2 = 0$ is equivalent to open-circuiting the input or output port. Hence, the z-parameters are called open-circuit impedance parameters. z_{11} is defined as the open-circuit input impedance, z_{22} is called the open-circuit output impedance and z_{12} and z_{21} are called the open-circuit transfer impedances. If $z_{12} = z_{21}$, the network is said to be reciprocal network. Also, if all the z-parameter are identical, then it is called asymmetrical network.

2. Admittance or Y-parameters

$$I_1 = Y_{11}V_1 + Y_{12}V_2 \quad ...(5)$$

$$I_2 = Y_{21}V_1 + Y_{22}V_2 \quad ...(6)$$

In matrix form,

$$[I]= [Y] [V] \quad ...(7)$$

$$\begin{bmatrix} I_1 \\ I_2 \end{bmatrix}=\begin{bmatrix} Y_{11} & Y_{12} \\ Y_{21} & Y_{22} \end{bmatrix}\begin{bmatrix} V_1 \\ V_2 \end{bmatrix} \quad ...(8)$$

$$y_{11}=\left.\frac{I_1}{V_1}\right|_{V_2=0}, \quad y_{21}=\left.\frac{I_2}{V_1}\right|_{V_2=0}$$

$$y_{12}=\left.\frac{I_1}{V_2}\right|_{V_1=0}, \quad y_{22}=\left.\frac{I_2}{V_2}\right|_{V_1=0}$$

3. Hybrid or H-parameters

$$V_1 = h_{11}I_1 + h_{12}V_2 \quad ...(9)$$

$$I_2 = h_{21}I_1 + h_{22}V_2 \quad ...(10)$$

In matrix form,

$$\begin{bmatrix} V_1 \\ I_2 \end{bmatrix} = \begin{bmatrix} h_{11} & h_{12} \\ h_{21} & h_{22} \end{bmatrix} \begin{bmatrix} I_1 \\ V_2 \end{bmatrix} \quad ...(11)$$

Solving these equations we get;

$$h_{11} \overset{\text{def}}{=} \left.\frac{V_1}{I_1}\right|_{V_2=0}, \quad h_{12} \overset{\text{def}}{=} \left.\frac{V_1}{V_2}\right|_{I_1=0}$$

$$h_{21} \overset{\text{def}}{=} \left.\frac{I_2}{I_1}\right|_{V_2=0}, \quad h_{22} \overset{\text{def}}{=} \left.\frac{I_2}{V_2}\right|_{I_1=0}$$

4. Transmission or ABCD Parameters

$$V_1 = AV_2 - B\,I_2 \quad ...(12)$$

$$I = CV_2 - DI_2 \quad ...(13)$$

$$\begin{bmatrix} V_1 \\ I_2 \end{bmatrix} = \begin{bmatrix} A & B \\ C & D \end{bmatrix} \begin{bmatrix} V_2 \\ -I_2 \end{bmatrix} \quad ...(14)$$

$$A = \left.\frac{V_1}{V_2}\right|_{I_2=0}, \quad B = \left.\frac{V_1}{-I_2}\right|_{V_2=0}, \quad C = \left.\frac{I_1}{V_2}\right|_{I_2=0}, \quad D = \left.\frac{I_1}{-I_2}\right|_{V_2=0}$$

These parameters can be measured under short or open circuit condition for use in the analysis of the circuit.

5.2 Transmission Line Parameters: H-parameters and Inverse H-parameters

A transmission line is used for the transmission of electrical power from generating substation to the various distribution units. It transmits the wave of voltage and current from one end to another. The transmission line is made up of a conductor having a uniform cross-section along the line. Air act as an insulating or dielectric medium between the conductors.

Transmission line.

For safety purpose, the distance between the line and ground is much more. The electrical tower is used for supporting the conductors of the transmission line. Towers are made up of steel for providing high strength to the conductor. For transmitting high voltage, over long distance high voltage direct current is used in the transmission line.

Parameters of Transmission Line

The performance of transmission line depends on the parameters of the line. The transmission line has mainly four parameters, resistance, inductance, capacitance and shunt conductance. These parameters are uniformly distributed along the line. Hence, it is also called the distributed parameter of the transmission line.

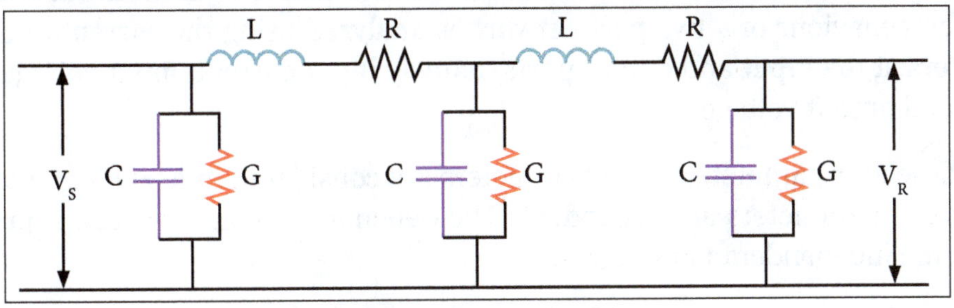

Transmission Line Model.

$$Z = R + j\omega L, \; Y = G = j\omega C$$

The inductance and resistance form series impedance whereas the capacitance and conductance form the shunt admittance. Some critical parameters of transmission line are explained below in detail,

- Line inductance: The current flow in the transmission line induces the magnetic flux. When the current in the transmission line changes, the magnetic flux also varies due to which emf induces in the circuit. The magnitude of inducing emf depends on the rate of change of flux. Emf produced in the transmission line resists the flow of current in the conductor and this parameter is known as the inductance of the line.

- Line capacitance: In the transmission lines, air acts as a dielectric medium. This dielectric medium constitutes the capacitor between the conductors, which store the electrical energy or increase the capacitance of the line. The capacitance of the conductor is defined as the presence of charge per unit of potential difference.

- Capacitance is negligible in short transmission lines whereas in long transmission, it is the most important parameter. It affects the efficiency, voltage regulation, power factor and stability of the system.

- Shunt conductance: Air act as a dielectric medium between the conductors. When the alternating voltage applies in a conductor, some current flow in the dielectric medium because of dielectric imperfections. Such current is called leakage current. Leakage current depends on the atmospheric condition and pollution like moisture and surface deposits.

Shunt conductance is defined as the flow of leakage current between the conductors. It is distributed uniformly along the whole length of the line. The symbol Y represented it and it is measured in Siemens.

5.2.1 Hybrid (H) Parameters or H-parameters

BJT is a two port device in which one terminal is common to both the input and output ports. The behaviour of a two port network is analyzed using the current and voltage parameters at the input and output ports, namely input current, input voltage, output current and output voltage.

Out of these four parameters, two parameters are considered independent and the remaining two parameters are dependent. The dependent parameters are expressed in terms of the independent parameters.

Let us consider the two port network shown in the below figure. The terminal behaviour of any two port network can be specified by the terminal voltages v_1 and v_2 at ports 1 and 2 respectively and currents i_1 and i_2 entering the ports 1 and 2 respectively.

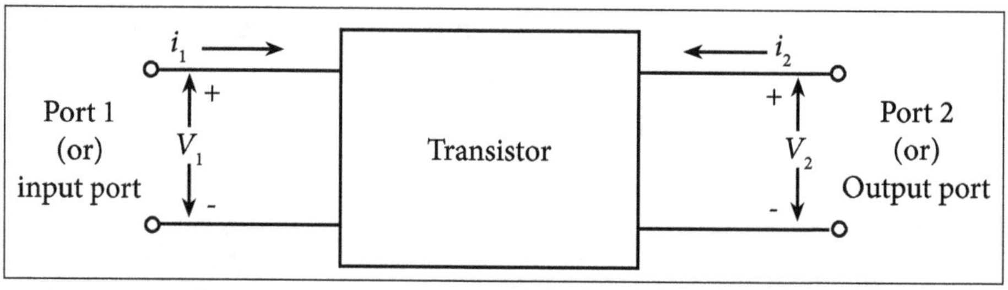

Two port network.

If the input current i_1 and the output voltage v_2 are taken as independent variables, the input voltage v_1 and output current i_2 can be written as,

$$v_1 = h_{11}i_1 + h_{12}v_2 \quad ...(1)$$

$$i_2 = h_{21}i_1 + h_{22}v_2 \quad ...(2)$$

The four hybrid parameters h_{11}, h_{12}, h_{21} and h_{22} are defined as follows:

- $h_{11} = \left[\frac{v_1}{i_1}\right]$ with $v_2 = 0$ which is input impedance with output port short circuited.
- $h_{22} = \left[\frac{i_2}{v_2}\right]$ with $i_1 = 0$ which is output admittance with input port open circuited.
- $h_{12} = \left[\frac{v_1}{v_2}\right]$ with $i_1 = 0$ which is reverse voltage transfer ratio with input port open circuited.
- $h_{21} = \left[\frac{i_2}{i_1}\right]$ with $v_2 = 0$ which is forward current gain with output port short circuited.

The dimensions of h-parameters are as follows,

$h_{11} - \Omega$

$h_{22} -$ mhos $(\mho)$

$h_{12}, h_{21} -$ Dimensionless

As the dimensions are not alike, (i.e.,) they are hybrid in nature, these parameters are called as hybrid parameters.

An alternative subscript notation recommended by IEEE is commonly used.

$i = 11 =$ Input

$o = 22 =$ Output

$f = 21 =$ Forward transfer

$r = 12 =$ Reverse transfer

Notations used in Transistor Circuit

In case of transistors, another subscript (b for base, e for emitter or c for collector) is added to indicate the type of configuration. For example,

$h_{11e} = h_{ie}$ = Short circuit input impedance

$h_{22e} = h_{oe}$ = Open circuit output admittance

$h_{12e} = h_{re}$ = Open circuit reverse voltage transfer ratio

$h_{21e} = h_{fe}$ = Short circuit forward current gain

The Hybrid Model for Two Port Network

Based on the definition of hybrid parameters, the mathematical model for two port networks known as h-parameter model can be developed as shown in the below figure.

The above equations (1) and (2) can be written as,

$$v_1 = h_i i_1 + h_r v_2$$

$$i_2 = h_f i_1 + h_o v_2$$

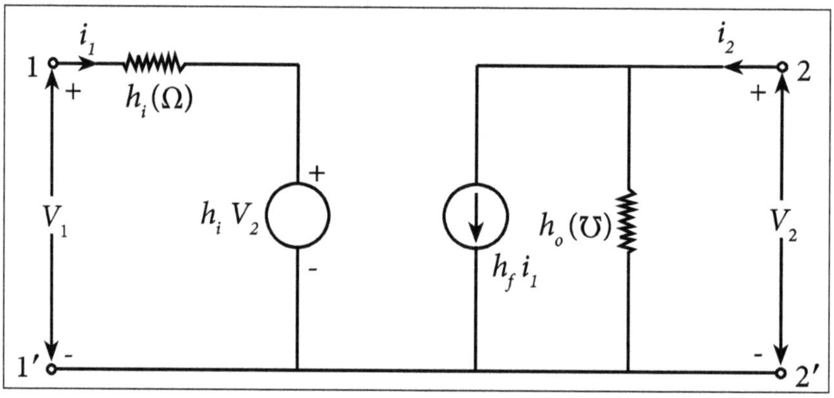

Hybrid model for two port network.

The model shown above must satisfy these two equations and it can be verified by writing Kirchhoff's voltage law equation for the input loop and KCL equation for the output node. The input circuit has a dependent voltage generator and the output circuit contains a dependent current generator.

H-parameter Model for CE Configuration

In case of common emitter (CE) configuration, the emitter of the transistor is common to both the input and output terminals. Base of the transistor is the input terminal and collector of the transistor is the output terminal. The variables are i_B, i_C, v_B $(= v_{BE})$

and $v_C (= V_{CE})$. i_B and v_C are considered as independent variables. v_B and i_C are considered as dependent variables.

$$v_B = f_1 (i_B, v_C)$$

$$i_C = f_2 (i_B, v_C)$$

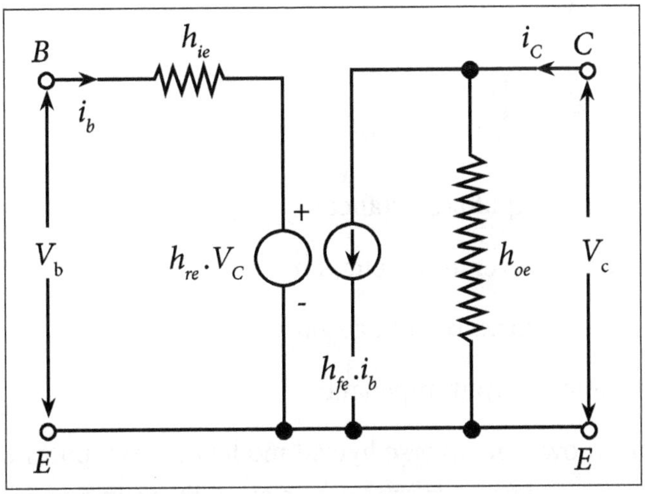

Hybrid model for transistor in CE configuration.

Therefore, the equations for CE configuration can be written as,

$$v_b = h_i e_{ib} + h_{re} v_C$$

$$i_c = h_{fe} i_b + h_{oe} v_C$$

The subscript 'e' indicates that, the h-parameters are for CE configuration.

h_{ie} = Input impedance when collector emitter terminal is short circuited.

h_{re} = Reverse voltage gain when base emitter terminal is open circuited.

h_{fe} = Forward current gain when collector emitter terminal is short circuited.

h_{oe} = Output admittance when base emitter terminal is open circuited.

5.2.2 Inverse H-parameters

A set of parameters closely related to the h parameters are the g parameters or inverse hybrid parameters. They are used to describe the terminal currents and voltages as,

$$I_1 = g_{11} V_1 + g_{12} I_2$$

$$V_2 = g_{21} V_1 + g_{22} I_2$$

$$\begin{bmatrix} I_1 \\ V_2 \end{bmatrix} = \begin{bmatrix} g_{11} & g_{12} \\ g_{21} & g_{22} \end{bmatrix} \begin{bmatrix} V_1 \\ I_2 \end{bmatrix} = [g] \begin{bmatrix} V_1 \\ I_2 \end{bmatrix}$$

The values of the g parameters are determined as,

$$g_{11} = \left.\frac{I_1}{V_1}\right|_{I_2=0}, \quad g_{12} = \left.\frac{I_1}{I_2}\right|_{V_1=0}$$

$$g_{21} = \left.\frac{V_2}{V_1}\right|_{I_2=0}, \quad g_{22} = \left.\frac{V_2}{I_2}\right|_{V_1=0}$$

g_{11} = Open-circuit input admittance

g_{12} = Short-circuit reverse current gain

g_{21} = Open-circuit forward voltage gain

g_{22} = Short-circuit output impedance

The following figure shows the inverse hybrid model of a two-port network.

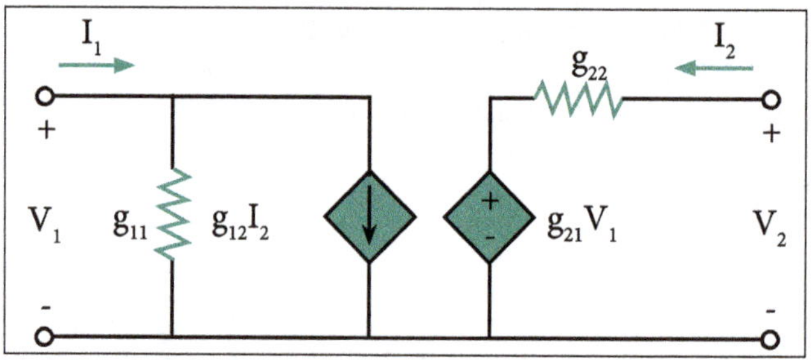

Inverse hybrid model.

5.3 Inverse Transmission Line Parameters and Relationship between Parameter Sets

b-parameters

The parameters b are called as the inverse transmission parameters,

$$b_{11} = \left.\frac{V_2}{V_1}\right|_{I_1=0}, \quad b_{12} = \left.\frac{V_2}{I_1}\right|_{V_1=0}$$

$$b_{21} = \left.\frac{I_2}{V_1}\right|_{I_1=0}, \quad b_{22} = -\left.\frac{I_2}{I_1}\right|_{V_1=0}$$

Where,

b_{11} = Open-circuit voltage gain

b_{12} = Negative short-circuit transfer impedance

b_{22} = Open circuit transfer admittance

b_{21} = Negative short-circuit current gain

5.3.1 Relationship between Parameter Sets

If all the two-port parameters for a network exist, it is possible to relate one set of parameters to another, since they interrelate the variables V_1, I_1, V_2 and I_2.

Relation between the z-parameters and y-parameters

The matrix equation for the z-parameters is expressed as,

$$\begin{bmatrix} V_1 \\ V_2 \end{bmatrix} = \begin{bmatrix} z_{11} & z_{12} \\ z_{21} & z_{22} \end{bmatrix} \begin{bmatrix} I_1 \\ I_2 \end{bmatrix}$$

$$V = zI \qquad \ldots(1)$$

Similarly, the equation for y-parameters is given by,

$$\begin{bmatrix} I_1 \\ I_2 \end{bmatrix} = \begin{bmatrix} y_{11} & y_{12} \\ y_{21} & y_{22} \end{bmatrix} \begin{bmatrix} V_1 \\ V_2 \end{bmatrix}$$

$$I = yV \qquad \ldots(2)$$

Substituting equation (2) into equation (1), we get

$$V = zy\,V$$

Hence,

$$z = y^{-1} = \frac{\text{adj}(y)}{\Delta y}$$

Where,

$$\Delta y = y_{11}\,y_{22} - y_{21}y_{12}$$

Thus we can obtain z-matrix by inverting y matrix. It is quite possible that a two-port network has a y-matrix or a z-matrix, but not both.

z-parameters in Terms of ABCD Parameters

The ABCD parameters of a two-port network are defined by,

$$V_1 = AV_2 - BI_2$$

$$I_1 = CV_2 - DI_2$$

$$V_2 = \frac{1}{C}(I_1 + DI_2)$$

$$V_2 = \frac{1}{C}I_1 + \frac{D}{C}I_2$$

$$V_1 = A\left(\frac{I_1}{C} + \frac{DI_2}{C}\right) - BI_2$$

$$= \frac{AI_1}{C} + \left(\frac{AD}{C} - B\right)I_2$$

By comparing the above equations with,

$$V_1 = z_{11}I_1 + z_{12}I_2$$

$$V_2 = z_{21}I_1 + z_{22}I_2$$

Respectively, we find that,

$$z_{11} = \frac{A}{C},\ z_{12} = \frac{AD - BC}{C},\ z_{21} = \frac{1}{C} \text{ and } z_{22} = \frac{D}{C}$$

Relation between Hybrid Parameters and z-parameters

$$V_1 = z_{11}I_1 + z_{12}I_2$$

$$V_2 = z_{21}I_1 + z_{22}I_2$$

From the above equation, we can write as below,

$$I_2 = \frac{-z_{22}}{z_{21}}I_2 + \frac{V_2}{z_{22}}$$

Substituting this value of I_2 in equation V_1, we get,

$$V_1 = z_{11}I_1 + z_{12}\left[\frac{-z_{21}I_1}{z_{22}} + \frac{V_2}{z_{22}}\right]$$

$$= \left[\frac{z_{11}z_{22} - z_{12}z_{21}}{z_{22}}\right]I_1 + \frac{z_{12}V_2}{z_{22}}$$

Comparing the above two equations with the following equations,

$$V_1 = h_{11}I_1 + h_{12}V_2$$

$$I_2 = h_{21}I_1 + h_{22}V_2$$

We get,

$$h_{11} = \frac{z_{11}z_{22} - z_{12}z_{21}}{z_{22}} = \frac{\Delta z}{Z_{22}}$$

$$h_{12} = \frac{z_{12}}{z_{22}},\ h_{21} = \frac{-z_{21}}{z_{22}},\ h_{22} = \frac{1}{z_{\rightarrow}}$$

Where,

$$\Delta z = z_{11}z_{22} - z_{12}z_{21}$$

Relationship between y-parameters and ABCD Parameters

$$I_1 = y_{11}V_1 + y_{12}V_2$$
$$I_2 = y_{21}V_1 + y_{22}V_2$$

From I_2 equation, we can write as,

$$V_1 = \frac{I_2}{y_{21}} - \frac{y_{22}}{y_{21}}V_2$$

$$= \frac{-y_{22}}{y_{21}}V_2 + \frac{1}{y_{21}}I_2$$

Substituting the value of V_1 in I_2, we get as,

$$I_1 = \frac{-y_{11}y_{22}}{y_{21}}V_2 + y_{12}V_2 + \frac{y_{11}}{y_{21}}I_2$$

$$= \frac{-\Delta y}{y_{21}}V_2 + \frac{y_{11}}{y_{21}}I_2$$

Comparing the above two equations with the following equations,

$$V_1 = AV_2 - BI_2$$
$$I_1 = CV_2 - DI_2$$

We get,

$$A = \frac{-y_{22}}{y_{21}}, B = \frac{-1}{y_{21}}, C = \frac{-\Delta y}{y_{21}}, D = \frac{-y_{11}}{y_{21}}$$

Where,

$$\Delta y = y_{11}y_{22} - y_{12}y_{21}$$

Parameter Relationships

	z	y	T	h
z	$\begin{bmatrix} z_{11} & z_{12} \\ z_{21} & z_{22} \end{bmatrix}$	$\begin{bmatrix} \frac{y_{22}}{\Delta y} & \frac{y_{12}}{\Delta y} \\ \frac{y_{21}}{\Delta y} & \frac{y_{11}}{\Delta y} \end{bmatrix}$	$\begin{bmatrix} \frac{A}{C} & \frac{\Delta T}{C} \\ \frac{1}{C} & \frac{D}{C} \end{bmatrix}$	$\begin{bmatrix} \frac{\Delta h}{h_{22}} & \frac{h_{12}}{h_{22}} \\ \frac{-h_{21}}{h_{22}} & \frac{1}{h_{22}} \end{bmatrix}$
y	$\begin{bmatrix} \frac{z_{22}}{\Delta z} & \frac{-z_{12}}{\Delta z} \\ \frac{-z_{21}}{\Delta z} & \frac{z_{11}}{\Delta z} \end{bmatrix}$	$\begin{bmatrix} y_{11} & y_{12} \\ y_{21} & y_{22} \end{bmatrix}$	$\begin{bmatrix} \frac{D}{B} & \frac{-\Delta T}{B} \\ \frac{-1}{B} & \frac{A}{B} \end{bmatrix}$	$\begin{bmatrix} \frac{1}{h_{11}} & \frac{-h_{12}}{h_{11}} \\ \frac{h_{21}}{h_{11}} & \frac{\Delta h}{h_{11}} \end{bmatrix}$
T	$\begin{bmatrix} \frac{z_{11}}{z_{21}} & \frac{\Delta z}{z_{21}} \\ \frac{1}{z_{21}} & \frac{z_{22}}{z_{21}} \end{bmatrix}$	$\begin{bmatrix} \frac{-y_{22}}{y_{21}} & \frac{-1}{y_{21}} \\ \frac{-\Delta y}{y_{21}} & \frac{-y_{11}}{y_{21}} \end{bmatrix}$	$\begin{bmatrix} A & B \\ C & D \end{bmatrix}$	$\begin{bmatrix} \frac{-\Delta h}{h_{21}} & \frac{-h_{11}}{h_{21}} \\ \frac{-h_{22}}{h_{21}} & \frac{-1}{h_{21}} \end{bmatrix}$
h	$\begin{bmatrix} \frac{\Delta z}{z_{22}} & \frac{z_{12}}{z_{22}} \\ \frac{-z_{21}}{z_{22}} & \frac{1}{z_{22}} \end{bmatrix}$	$\begin{bmatrix} \frac{1}{y_{11}} & \frac{-y_{12}}{y_{11}} \\ \frac{y_{21}}{y_{11}} & \frac{\Delta y}{y_{11}} \end{bmatrix}$	$\begin{bmatrix} \frac{B}{D} & \frac{\Delta T}{D} \\ \frac{-1}{D} & \frac{C}{D} \end{bmatrix}$	$\begin{bmatrix} h_{11} & h_{12} \\ h_{21} & h_{22} \end{bmatrix}$
$\Delta z = z_{11}z_{22} - z_{12}z_{21}$, $\Delta y = y_{11}y_{22} - y_{12}y_{21}$, $\Delta h = h_{11}h_{22} - h_{12}h_{21}$, $\Delta T = AD - BC$				

5.4 Parallel Connection, Cascading and Series Connection of Two Port Networks

The very purpose of defining and evaluating the five different sets of describing parameter sets for a linear time-invariant two-port network lies in the relative ease of combining them for various types of interconnections between two two-port networks.

There are five different ways of interconnecting two linear time-invariant two-port networks. The first interconnection is the cascade connection in which the second port of the first two-port network feeds the first port of the second two-port network. This interconnection is shown in the below figure.

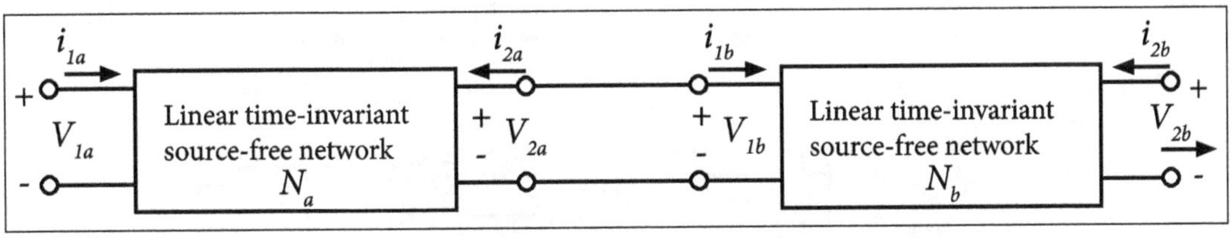

Cascade connection of two port networks.

We show that the ABCD matrix of the overall two-port network is the product of ABCD matrices of N_a and N_b.

$$\begin{bmatrix} v_{1b} \\ i_{1b} \end{bmatrix} = \begin{bmatrix} A_b & B_b \\ C_b & D_b \end{bmatrix} \begin{bmatrix} v_{2b} \\ -i_{2b} \end{bmatrix}$$

$$\therefore \quad \begin{bmatrix} v_{1a} \\ i_{1a} \end{bmatrix} = \begin{bmatrix} A_a & B_a \\ C_a & D_a \end{bmatrix} \begin{bmatrix} v_{2a} \\ -i_{2a} \end{bmatrix} = \begin{bmatrix} A_a & B_a \\ C_a & D_a \end{bmatrix} \begin{bmatrix} A_b & B_b \\ C_b & D_b \end{bmatrix} \begin{bmatrix} v_{2b} \\ -i_{2b} \end{bmatrix}$$

ABCD matrix of the overall two-port $= \begin{bmatrix} A & B \\ C & D \end{bmatrix} = \begin{bmatrix} A_a & B_a \\ C_a & D_a \end{bmatrix} = \begin{bmatrix} A_b & B_b \\ C_b & D_b \end{bmatrix}$.

This result can be extended to n networks connected in cascade to arrive at the general result,

$$\begin{bmatrix} A & B \\ C & D \end{bmatrix} = \begin{bmatrix} A_1 & B_1 \\ C_1 & D_1 \end{bmatrix} \begin{bmatrix} A_2 & B_2 \\ C_2 & D_2 \end{bmatrix} \cdots \begin{bmatrix} A_n & B_n \\ C_n & D_n \end{bmatrix}$$

No other parameter set (i.e., y, z, h or g) will combine so elegantly for a cascade connection. Obviously, transmission parameter based description is the most convenient description for a two-port network in cascading context.

Note that N_B followed by N_A is not the same as N_A followed by N_B — the ordering of cascade matters. Matrix product is not commutative. The remaining four interconnections are shown in the below figure.

The connection in figure (a) is termed Series Connection and in this connection both the input and the output ports are connected in series. It can be shown that the z-parameters of the overall two-port network is the sum of corresponding z-parameters of N_a and N_b.

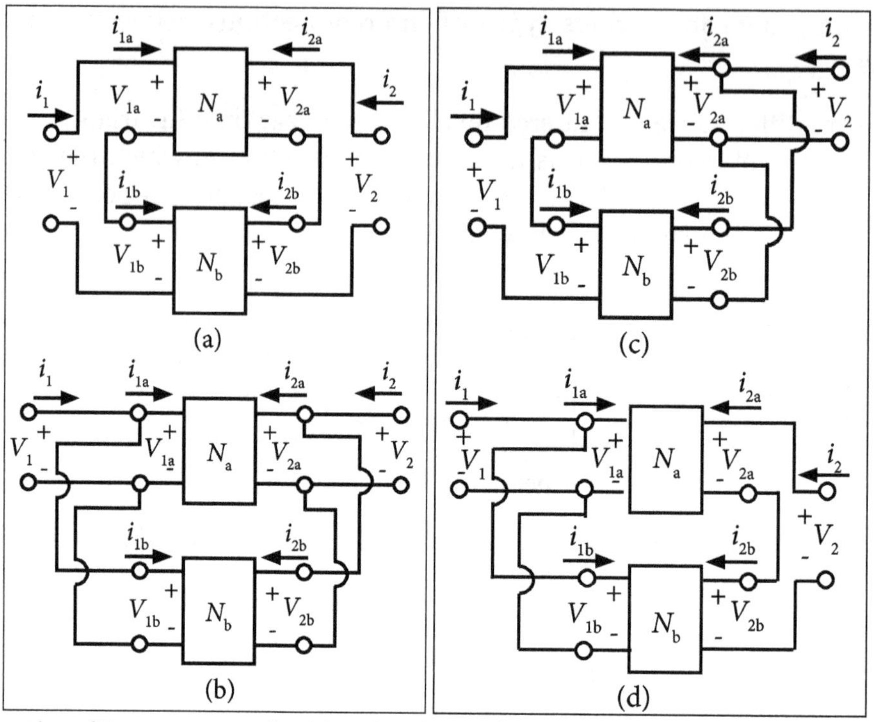

Interconnection of Two port networks: (a) Series, (b) Parallel, (c) Series-Parallel and (d) Parallel-Series.

The connection in figure (b) is called as Parallel Connection. Both the input ports and the output ports are tied in parallel in this connection. The y-parameters get added in this case. The connection in figure(c) is a Series-Parallel Connection in which the input ports are connected in series and the output ports are connected in parallel. The h-parameters get added in this case.

The connection in figure (d) is called as Parallel-Series Connection. The input ports are in parallel and the output ports are in series in this connection. The g-parameters get added in this case. These four connections appear in the context of analysis of negative feedback amplifiers in the area of Analogue Electronic Circuits. The various two-port interconnections and the interconnection equations are summarized in table.

Table: Two-port interconnections and interconnection equations.

Connection name	Inter connection equation
Cascade	$\begin{bmatrix} A & B \\ C & D \end{bmatrix} \begin{bmatrix} A_a & B_a \\ C_a & D_a \end{bmatrix} \begin{bmatrix} A_b & B_b \\ C_b & D_b \end{bmatrix}$
Series	$\begin{bmatrix} z_{11} & z_{12} \\ z_{21} & z_{22} \end{bmatrix} = \begin{bmatrix} z_{11a} & z_{12a} \\ z_{21a} & z_{22a} \end{bmatrix} + \begin{bmatrix} z_{11b} & z_{12b} \\ z_{21b} & z_{22b} \end{bmatrix}$
parallel	$\begin{bmatrix} y_{11} & y_{12} \\ y_{21} & y_{22} \end{bmatrix} = \begin{bmatrix} y_{11a} & y_{12a} \\ y_{21a} & y_{22a} \end{bmatrix} + \begin{bmatrix} y_{11b} & y_{12b} \\ y_{21b} & y_{22b} \end{bmatrix}$

Series –Parallel	$\begin{bmatrix} h_{11} & h_{12} \\ h_{21} & h_{22} \end{bmatrix} = \begin{bmatrix} h_{11a} & h_{12a} \\ h_{21a} & h_{22a} \end{bmatrix} + \begin{bmatrix} h_{11b} & h_{12b} \\ h_{21b} & h_{22b} \end{bmatrix}$
Parallel – series	$\begin{bmatrix} g_{11} & g_{12} \\ g_{21} & g_{22} \end{bmatrix} = \begin{bmatrix} g_{11a} & g_{12a} \\ g_{21a} & g_{22a} \end{bmatrix} + \begin{bmatrix} g_{11b} & g_{12b} \\ g_{21b} & g_{22b} \end{bmatrix}$

5.4.1 Problem Solving Including Dependent Sources

1. Let us determine the admittance parameters of the T network shown in the below figure.

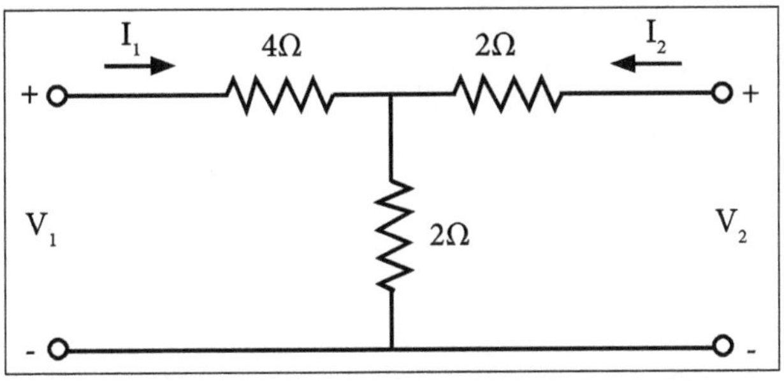

Solution:

Given:

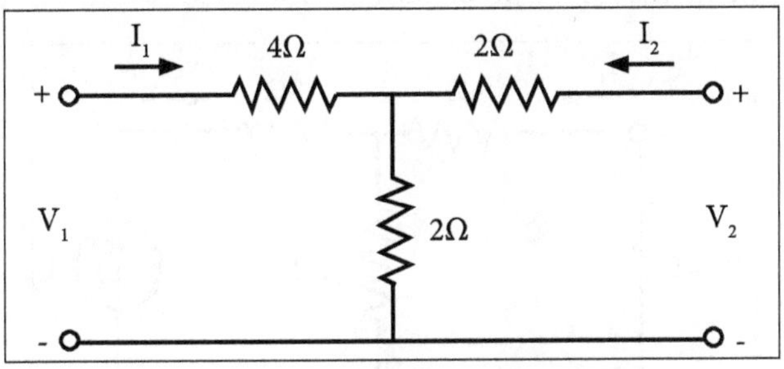

To find y_{11} and y_{21}, we have to short the output terminals and connect a current source I_1 to the input terminals. The circuit so obtained is shown in the given figure (a).

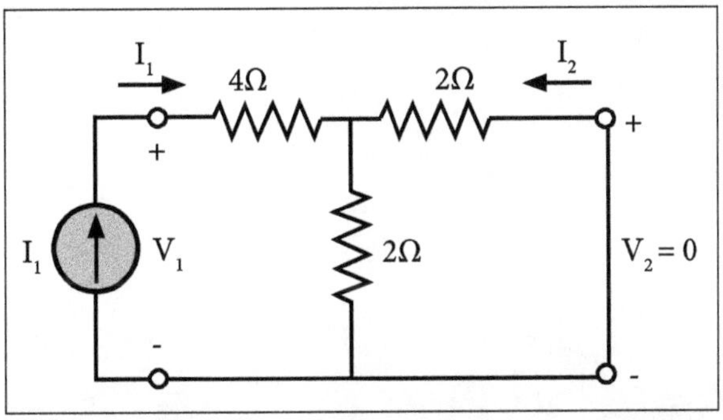

(a)

$$I_1 = \frac{V_1}{4+\frac{2\times 2}{2+2}} = \frac{V_1}{5}$$

Here,

$$y_{11} = \left.\frac{I_1}{V_1}\right|_{V_2=0} = \frac{1}{5}S$$

Using the principle of current division,

$$-I_2 = \frac{I_1 \times 2}{2+2} = \frac{I_1}{2}$$

$$\Rightarrow \qquad -I_2 = \frac{1}{2}\left[\frac{V_1}{5}\right]$$

Hence,

$$y_{21} = \left.\frac{I_2}{V_1}\right|_{V_2=0} = \frac{-1}{10}S$$

To find y_{12} and y_{22}, we have to short-circuit the input terminals and connect a current source I_2 to the output terminals. The circuit so obtained is shown in below figure (b).

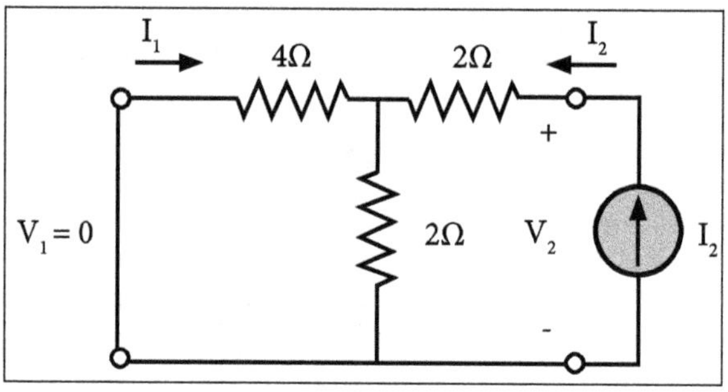

(b)

$$I_2 = \frac{V_2}{2+\frac{4\times 2}{4+2}}$$

$$= \frac{V_2}{2+\frac{4}{3}}$$

$$= \frac{3V_2}{10}$$

Hence,

$$y_{22} = \left.\frac{I_2}{V_2}\right|_{V_1=0} = \frac{3}{10}S$$

Employing the principle of current division, we have

$$-I_1 = \frac{I_2 \times 2}{2+4}$$

$$\Rightarrow \qquad -I_1 = \frac{2I_2}{6}$$

$$\Rightarrow \qquad -I_1 = \frac{1}{3}\left[\frac{3V_2}{10}\right]$$

Hence,

$$y_{12} = \left.\frac{I_1}{V_2}\right|_{V_1=0} = \frac{-1}{10}S$$

It may be noted that, $y_{12} = y_{21}$.

Thus, in matrix form we have

$$I = YV$$

$$\Rightarrow \qquad \begin{bmatrix} I_1 \\ I_2 \end{bmatrix} = \begin{bmatrix} \frac{1}{5} & \frac{-1}{10} \\ \frac{-1}{10} & \frac{3}{10} \end{bmatrix} \begin{bmatrix} V_1 \\ V_2 \end{bmatrix}$$

2. Here we shall determine the z parameters for the two port network shown in the given figure.

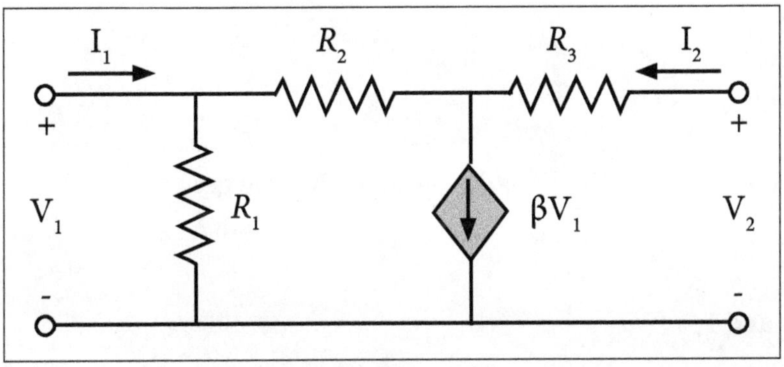

Solution:

Given:

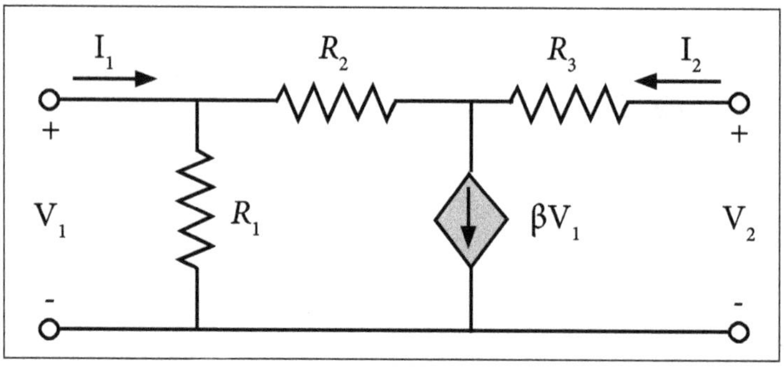

To find z_{11} and z_{21}, the output terminals are open-circuited and a voltage source is connected to the input terminals. The resulting circuit is shown in figure (a).

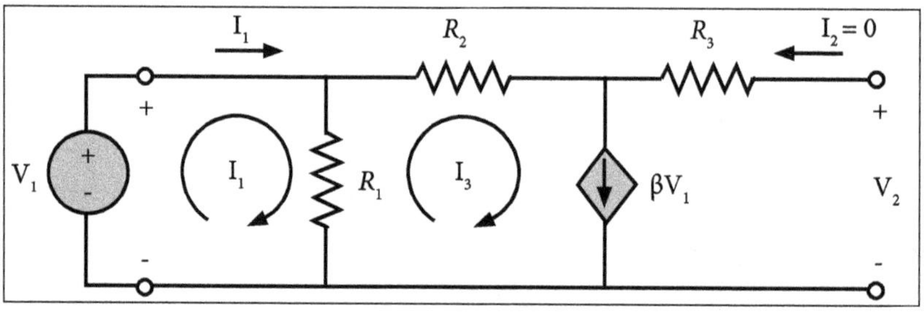

(a)

By inspection, we find that $I_3 = \beta V_1$

$$R_1(I_1 - I_3) = V_1$$

$$\Rightarrow \quad R_1 I_1 - R_1 I_3 = V_1$$

$$\Rightarrow \quad R_1 I_1 - R_1 \beta V_1 = V_1$$

$$\Rightarrow \quad (1 + R_1 \beta) V_1 = R_1 I_1$$

Hence,

$$z_{11} = \left.\frac{V_1}{I_1}\right|_{I_2=0} = \frac{R_1}{1+\beta R_1}$$

By applying KVL to the path, $V_1 \to R_2 \to R_3 \to V_2$, we get

$$-V_1 + R_2 I_3 - R_3 I_2 + V_2 = 0$$

Since $I_2 = 0$ and $I_3 = \beta V_1$,

We get,

$$-V_1 + R_2\beta V_1 - 0 + V_2 = 0$$

$$\Rightarrow \qquad V_2 = V_1(1 - \beta R_2)$$

$$= (1-\beta R_2)\frac{R_1 I_1}{1+\beta R_1}$$

Hence,

$$z_{21} = \left.\frac{V_2}{I_1}\right|_{I_2=0} = \frac{R_1(1-\beta R_2)}{1+\beta R_1}$$

The circuit used for finding z_{12} and z_{22} is shown in figure (b).

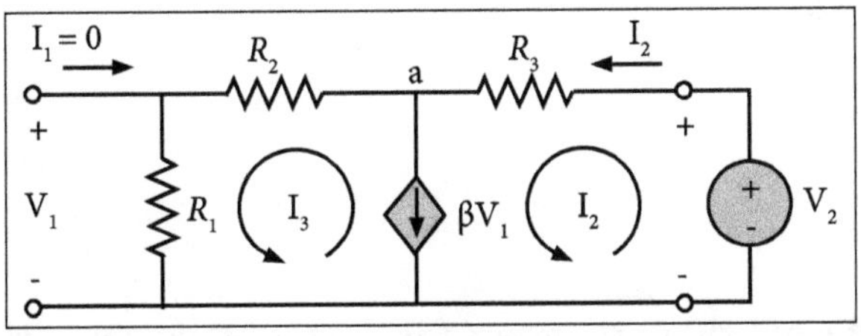

(b)

By inspection, we find that,

$$V_1 = R_1 I_1 + R_3(I_1 + I_2)$$

$$= (R_1 + R_3)I_1 + R_3 I_2$$

Applying KVL to the path $R_3 \rightarrow R_1 \rightarrow V_2$, we get

$$R_3 I_3 + (R_2 + R_1)I_3 - V_2 = 0$$

$$\Rightarrow \qquad R_3 I_2 + (R_2 + R_1)\frac{I_2}{1+\beta R_1} = V_2$$

$$\Rightarrow \qquad I_2\left[R_3 + \frac{R_2 + R_1}{1+\beta R_1}\right] = V_2$$

Hence,

$$z_{22} = \left.\frac{V_2}{I_2}\right|_{I_1=0}$$

$$=R_3+\frac{R_2+R_1}{1+\beta R_1}\Omega$$

Applying KCL at node a, we get,

$$\beta V_1+I_3=I_2$$

$$\Rightarrow \quad \beta V+\frac{V_1}{R_1}=I_2$$

$$\Rightarrow \quad V_1\left(\beta+\frac{1}{R_1}\right)=I_2$$

$$\Rightarrow \quad Z_{12}=\left.\frac{V_1}{I_2}\right|_{I_1=0}=\frac{1}{\beta+\frac{1}{R_1}}$$

$$=\frac{R_1}{1+\beta R_1}$$

3. Let us determine the hybrid parameters for the two-port network shown in the below figure (a).

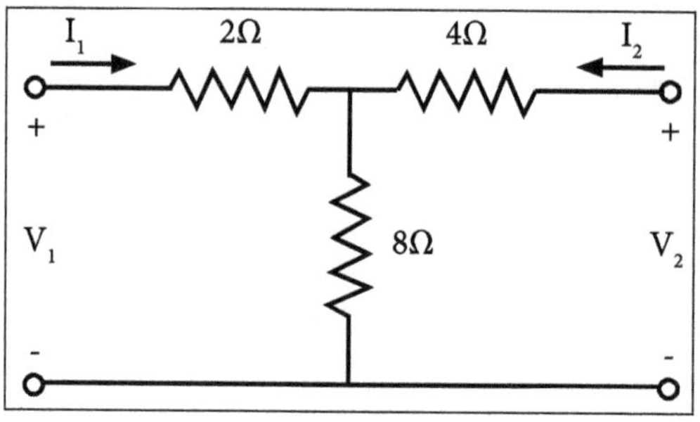

(a)

Solution:

Given:

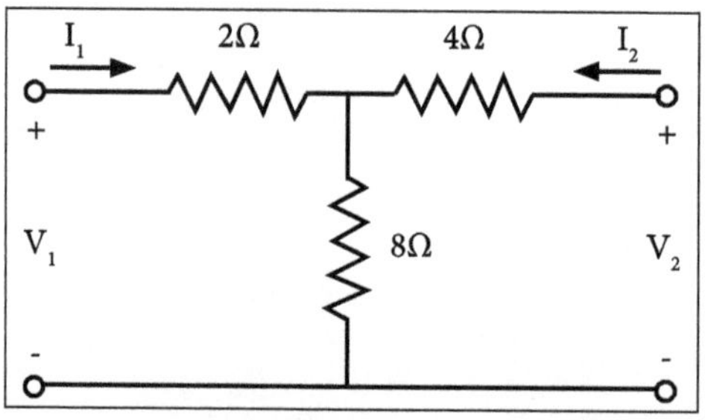

To find h_{11} and h_{21}, short-circuit the output port and connect a current source I_1 to the input port as shown in figure (b).

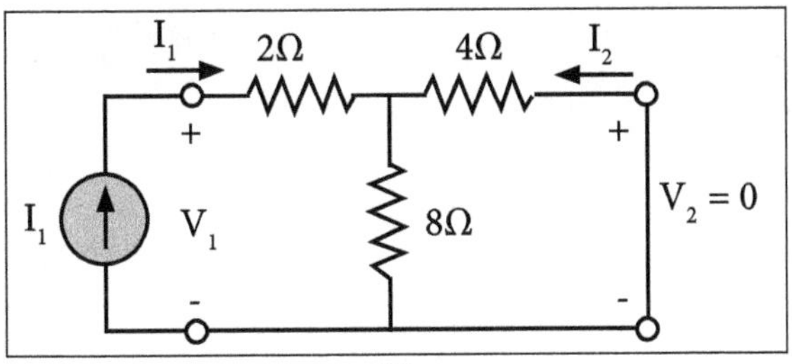

(b)

Referring to figure (b), we find that,

$$V_1 = I_1\left[2\Omega + \left(8\Omega \| 4\Omega\right)\right]$$

$$= I_1 \times 4.67$$

Hence,

$$h_{11} = \left.\frac{V_1}{I_2}\right|_{V_2=0} = 4.67\Omega$$

By using the principle of current division, we find that,

$$-I_2 = \frac{I_1 \times 8}{8+4} = \frac{2}{3} I_1$$

Hence,

$$h_{21} = \left.\frac{I_2}{I_1}\right|_{V_2=0} = \frac{-2}{3}$$

To obtain h_{12} and h_{22}, open-circuit the input port and connect a voltage source V_2 to the output port as in figure(c).

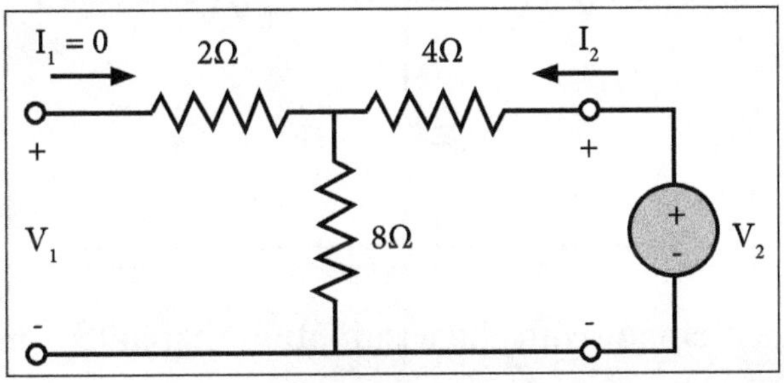

(c)

Using the principle of voltage division,

$$V_1 = \frac{8}{8+4} V_2 = \frac{2}{3} V_2$$

Hence,

$$h_{12} = \frac{V_1}{V_2} = \frac{2}{3}$$

Also,

$$V_2 = (8+4) I_2$$

$$= 12 I_2$$

$$\Rightarrow \qquad h_{22} = \left. \frac{I_2}{V_2} \right|_{I_2 = 0} = \frac{1}{12} S$$

4. Let us determine the transmission parameters in the s domain for the network shown in the below figure.

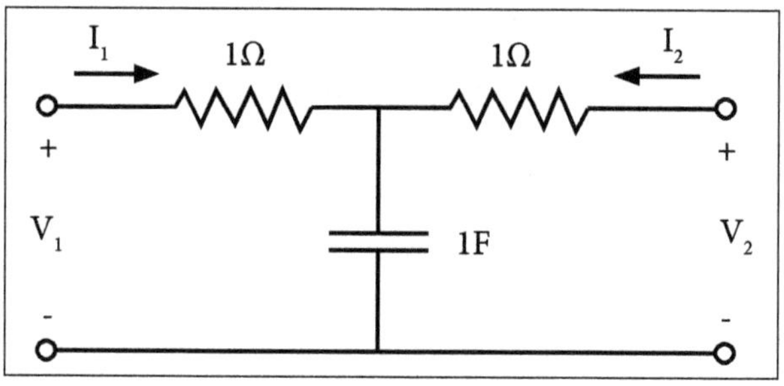

Solution

Given:

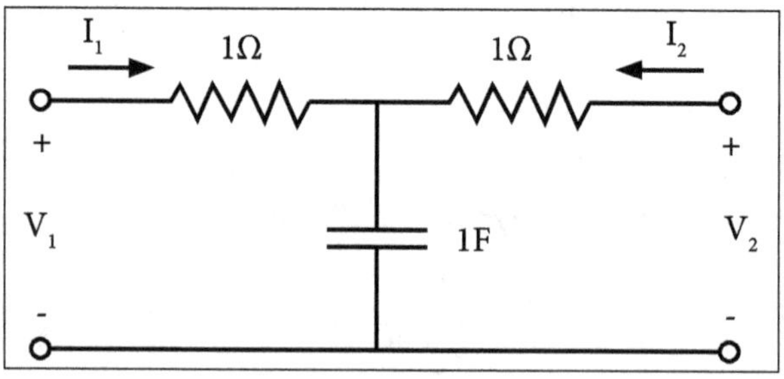

The domain equivalent circuit with the assumption that all the initial conditions are zero is shown in figure (a).

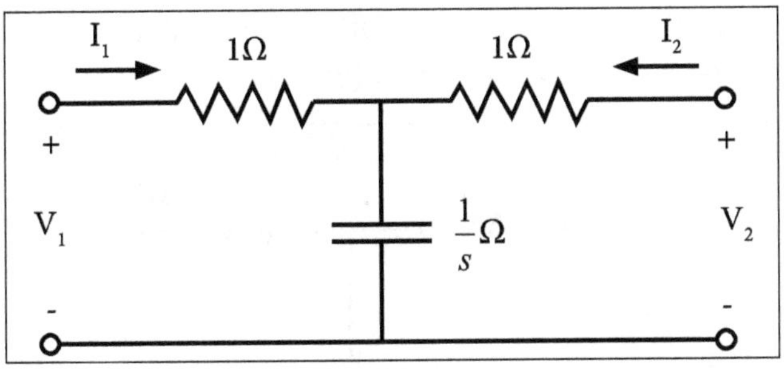

(a)

To find the parameters A and C, open-circuit the output port and connect a voltage source V_1 at the input port. The same is shown in figure (b).

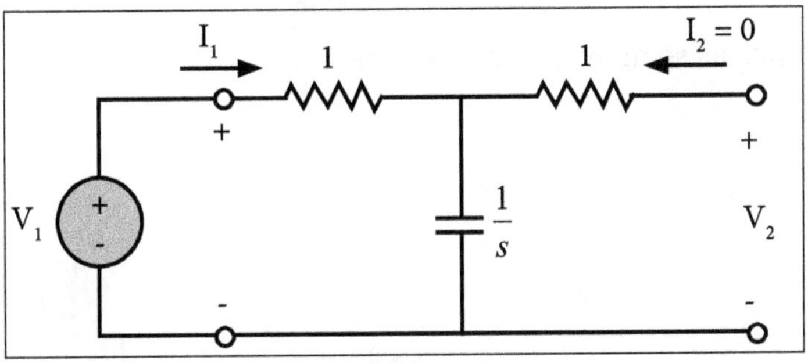

(b)

$$I_1 = \frac{V_1}{1+\frac{1}{s}} = \frac{sV_1}{s+1}$$

Then,

$$V_2 = \frac{1}{s} I_1$$

$$\Rightarrow \quad V_2 = \frac{1}{s}\frac{sV_1}{s+1} = \frac{V_1}{s+1}$$

$$\Rightarrow \quad A = \left.\frac{V_1}{V_2}\right|_{I_2=0} = s+1$$

Also,

$$V_2 = \frac{1}{s} I_1$$

$$\Rightarrow \quad C = \left.\frac{I_1}{V_2}\right|_{I_2=0} = s$$

To find the parameters B and D, short-circuit the output port and connect a voltage source V_1 to the input port as shown in figure(c).

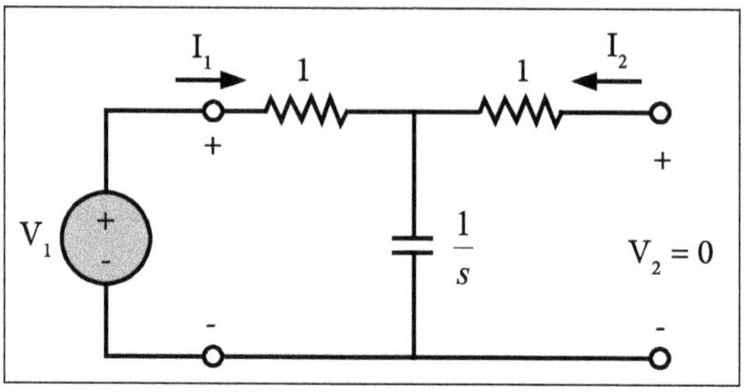

(c)

The total impedance as seen by the source V_1 is,

$$Z=1+\frac{\frac{1}{s}\times 1}{\frac{1}{s}+1}$$

$$=1+\frac{1}{s+1}=\frac{s+2}{s+1}$$

$$I_1=\frac{V_1}{Z}=\frac{V_1(s+1)}{(s+2)} \qquad ...(1)$$

Using the principle of current division, we have,

$$-I_2=\frac{I_1\left(\frac{1}{s}\right)}{\frac{1}{s}+1}=\frac{I_1}{s+1} \qquad ...(2)$$

Hence,

$$D=\left.\frac{I_1}{-I_2}\right|_{V_2=0}=s+1$$

From equation (1) and (2), we can write,

$$-I_2(s+1)=\frac{V_1(s+1)}{(s+2)}$$

Hence,

$$B = \left.\frac{-V_1}{I_2}\right|_{V_2=0} = s+2$$

Verification

We know that for a two port network without any dependent sources, $AD - BC = 1$

$$(s+1)(s+1) - s(s+2) = 1.$$

Hence, verified.

6

Transients

6.1 RL Circuit, RC Circuit with DC Excitation and Evaluating Initial Conditions Procedure

A transient response or natural response is the response of a system to a change from equilibrium. The transient response is not necessarily tied to "on/off" events but to any event that affects the equilibrium of the system.

First Order Differential Equations of Transient Response

First-order RC Circuit

Let us consider a very simple dynamic circuit, which contains one capacitor.

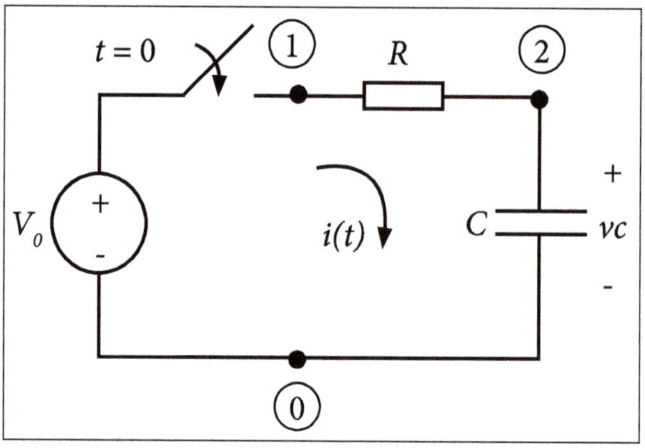

RC circuit.

After t = 0, the circuit is closed. So, we can easily write as,

$$i(t) = \frac{vR}{R} = C\frac{dv_C}{dt}$$

And,

$$vR(t) = v_1(t) - v_C(t)$$

Thus, we have,

$$\frac{v_1(t)-v_C(t)}{CR}=\frac{dv_C}{dt} \Rightarrow \frac{dv_C}{dt}+\frac{v_C}{CR}=\frac{v_1(t)}{CR} \qquad \therefore V_o \text{ for } t>0$$

Thus, we have,

$$v_c(t)=Ae^{-\frac{t}{CR}}+V_o \qquad \text{for } t\geq 0$$

If the initial condition is $V_C(0^+)=0$, then $A=-V_0$.

Thus, the solution is,

$$v_c(t)=V_o\left(1-e^{-\frac{t}{CR}}\right)$$

Transient Response of the RC Circuit

Once we have the capacitor voltage, we can find anything.

Starting with $v_c(t)=V_o\left(1-e^{-\frac{t}{CR}}\right)$

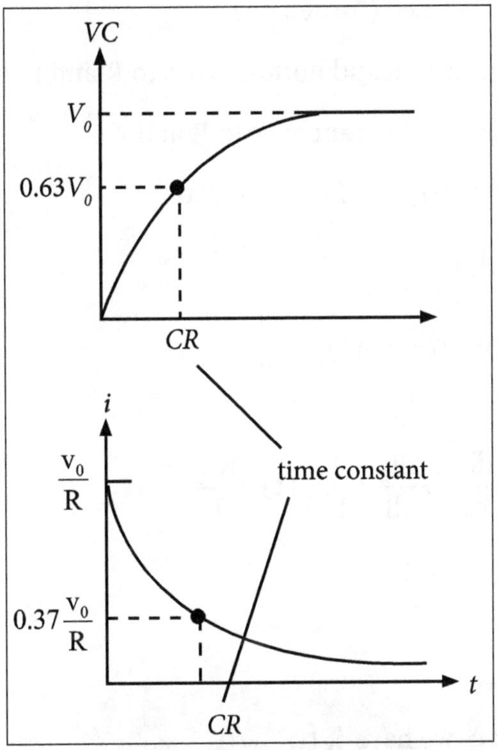

Transient response of RC circuit.

We can derive the current as,

$$i(t)=C\frac{dv_C}{dt}=\frac{V_O}{R}e^{-\frac{t}{CR}}$$

We see the solution typically has a Transient which dies out eventually and as t tends to ∞, the solution settles to a steady state.

First-order RL Circuit

Let us consider a below RL circuit,

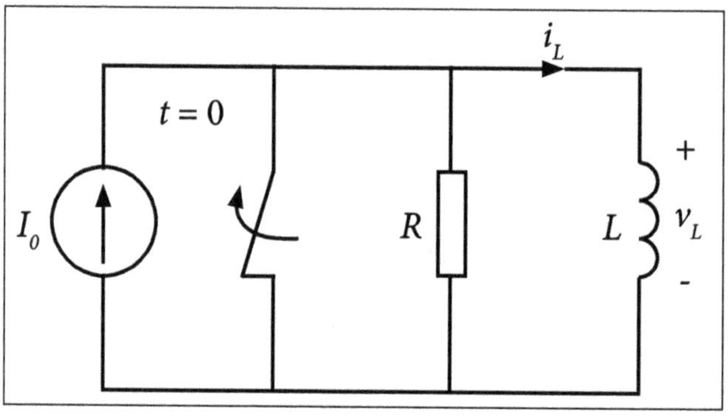

RL Circuit.

Before t = 0, the switch is closed (turned on).

Current goes through the switch and nothing goes to R and L. Initially, $i_L(0^-)=0$.

At t = 0, the switch is opened. Current goes to R and L.

We know from KCL that $I_0 = i_R + i_L$ for $t > 0$, i.e.,

$$v_R(t)=(I_0-i_L(t))R$$

The constitutive relations give $v_L(t)=L\frac{di_L}{dt}$

Hence,

$$(I_0-i_L(t))R=L\frac{di_L}{dt}\Rightarrow\frac{di_L}{dt}+\frac{R}{L}i_L(t)=\frac{RI_O}{L}$$

The solution is given as,

$$i_L(t)=Ae^{\frac{-Rt}{L}}+I_0$$

From the initial condition, we have $i_L(0^-)=0$

Continuity of the inductor current means that $i_L(0^+) = i_L(0^-) = 0$

Hence,

$$A = -I_o$$

Thus,

$$i_L(t) = I_o\left(1 - e^{\frac{-Rt}{L}}\right)$$

Transient Response of the RL Circuit

Starting with $i_L(t) = I_o\left(1 - e^{\frac{-Rt}{L}}\right)$

We can find $v_L(t)$:

$$v_L(t) = RI_o e^{\frac{-Rt}{L}}$$

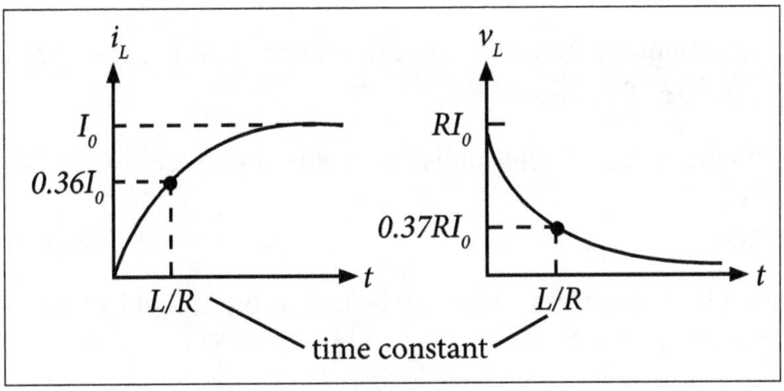

Transient response of RL circuit.

First order transients are always like these,

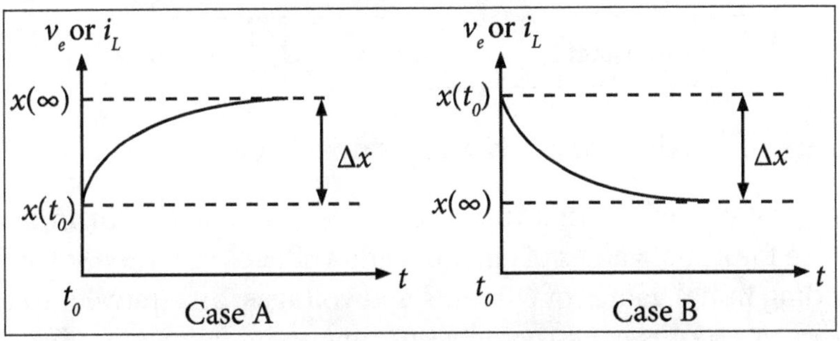

1st order transient response.

Case A: $x(t)=x(t_o)+\Delta x\left(1-e^{-(t-t_o/\tau)}\right)$

Case B: $x(t)=x(t_o)+\Delta x\left(1-e^{-(t-t_o/\tau)}\right)$

Transient Time

The time taken for a circuit to change from one steady state to another steady state is called transient time.

Free and Forced Responses

The response of a circuit depends upon the nature of the circuit is called free or natural response.

The response of a circuit depends upon the source of the circuit is called forced response.

In general, any circuit response classified into,

$I(t)=i_c+i_p$ [or] complementary function + particular integral

Where,

i_c – Complementary function dies after some time is called natural (or) transient (or) source free response.

i_p – Particular integral solution is the steady state response or forced response.

Time Constant

The time required for the current to rise to the 0.632 (63.2%) of its final value is known as time constant of the circuit.

Expression of time Constants of Series RL and RC Circuit

Circuit	RL circuit	RC circuit
Time constant	$\tau=L/R$ sec	$\tau=RC$ sec

6.1.1 Evaluating Initial Conditions Procedure

There is no unique procedure that must be followed in solving for initial conditions. We usually solve for initial values of currents and voltages and then solve for the derivatives. For finding initial values of currents and voltages, an equivalent network of the original network at $t = 0^+$ is constructed according to the following rules:

- Replace all the inductors with the open circuit or with the current sources having the value of current flowing at $t = 0^{+}$.
- Replace all the capacitors with the short circuits or with the voltage source of value $v_o = \frac{qo}{C}$ if there is the initial charge.
- Resistors are left in the network without any changes.

6.2 Second Order Differential Equations, Homogeneous and Non-homogenous Equations

A second order ordinary differential equation has the general form,

$$y'' = f(t, y, y')$$

Where, f is some given function.

This equation is said to be linear if f is linear in y and y',

$$y'' = g(t) - p(t)y' - q(t)y$$

Otherwise the equation is said to be nonlinear.

A second order linear equation often obtained as,

$$p(t)y'' - Q(t)y' + R(t)y = G(t)$$

If $G(t) = 0$ for all t, then the equation is called homogeneous. Otherwise, the equation is non-homogeneous.

Generation of a Differential Equation in an Output Variable

The output equation $y = Cx + Du$ for any system variable is a single algebraic equation,

$$y(t) = \begin{bmatrix} c_1 & c_2 \end{bmatrix} \begin{bmatrix} x_1 \\ x_2 \end{bmatrix} + [d]u(t)$$

$$= c_1 x_1(t) + c_2 x_2(t) + du(t) \qquad \text{...(1)}$$

And in the Laplace domain,

$$Y(s) = \left(C(sI - A)^{-1} B + D\right) U(s)$$

$$=\frac{1}{\det[sI-A]}\left(C\,adj(sI-A)+\det[sI-A]D\right)$$

The determinants may be expanded and the resulting equation written as a differential equation:

$$\frac{d^2y}{dt^2}-(a_{11}+a_{22})\frac{dy}{dt}+(a_{11}a_{22}-a_{12}a_{21})y=q_2\frac{d^2u}{dt^2}+q_1\frac{du}{dt}+q_0u \qquad ...(2)$$

or in terms of the standard system parameters,

$$\frac{d^2y}{dt^2}+2\zeta\omega n\frac{dy}{dt}+\omega_n^2y=q_2\frac{d^2u}{dt^2}+q_1\frac{du}{dt}+q_0u \qquad ...(3)$$

Where the coefficient q_0, q_1, and q_2 are,

$$q_0=c_1(-a_{22}b_1+a_{12}b_2)+c_2(-a_{11}b_2+a_{21}b_1)+d(a_{11}a_{22}-a_{12}a_{21})$$

$$q_1=c_1b_1+c_2b_2-d(a_{11}+a_{22})$$

$$q_2=d \qquad ...(4)$$

Notice that the left hand side of the differential equation is the same for all system variables and that the only difference between any of the differential equations describing any system variable is in the constant coefficients q_2, q_1 and q_0 on the right hand side.

Solution of the Homogeneous Second-order Equation

For any system variable y(t) in a second-order system, the homogeneous equation is found by setting the input u(t) = 0 so that equation(3) becomes,

$$\frac{d^2y}{dt^2}+2\zeta\omega n\frac{dy}{dt}+\omega_n^2y=0 \qquad ...(5)$$

The solution, $y_h(t)$, to the homogeneous equation is found by assuming the general exponential form,

$$yh(t)=C_1e^{\lambda_1t}+C_2e^{\lambda_2t} \qquad ...(6)$$

Where C_1 and C_2 are constants defined by the initial conditions and the eigenvalues λ_1 and λ_2 are the roots of the characteristic equation.

$$\det[sI-A]=\lambda^2+2\zeta\omega_n\lambda+\omega_n^2=0 \qquad ...(7)$$

Found using the quadratic formula,

$$\lambda_1,\lambda_2=-\zeta\omega_n \pm \omega_n\sqrt{\zeta^2-1} \qquad ...(8)$$

If $\zeta=1$, the two roots are equal $(\lambda_1=\lambda_2=\lambda)$, a modified form for the homogeneous solution is necessary,

$$yc(t)=C_1e^{\lambda t}+C_2e^{\lambda t} \qquad ...(9)$$

In either case, the homogeneous solution consists of two independent exponential components, with two arbitrary constants, C_1 and C_2, whose values are selected to make the solution satisfy a given pair of initial conditions.

The initial conditions for the output variable may be specified directly as part of the problem statement or they may have to be determined from knowledge of the state variables $x_1(0)$ and $x_2(0)$ at time t = 0. The homogeneous output equation may be used to compute y(0) directly from elements of the A and C matrices,

$$y(0)=c_1x_1(0)+c_2x_2(0) \qquad ...(10)$$

and the value of the derivative $\dot{y}(0)$ may be determined by differentiating the output equation and substituting for the derivatives of the state variables from the state equations,

$$\dot{y}(0)=c_1\dot{x}_1(0)+c_2\dot{x}_2(0)$$

$$=c_1\left(a_{11}x_1(0)+a_{12}x_2(0)\right)+c_2\left(a_{21}x_1(0)+a_{22}x_2(0)\right) \qquad ...(11)$$

To illustrate the influence of damping ratio and natural frequency on the system response, we consider the response of an unforced system output variable with initial output conditions of $y(0)=y_0$ and $\dot{y}(0)=0$ If the roots of the characteristic equation are distinct, imposing these initial conditions on the general solution of equation(6) gives,

$$y(0)=y_0=C_1+C_2$$

$$\left.\frac{dy}{dt}\right|_{t=0}=0=\lambda_1C_1+\lambda_2C_2 \qquad ...(12)$$

With the result that,

$$C_1=\frac{\lambda_2}{\lambda_2-\lambda_1}y_0 \quad \text{and} \quad C_2=\frac{\lambda_1}{\lambda_1-\lambda_2}y_0 \qquad ...(13)$$

For this set of initial conditions, the homogeneous solution is therefore,

$$y_h(t) = y_o\left[\left(\frac{\lambda_2}{\lambda_2-\lambda_1}\right)e^{\lambda_1 t} + \left(\frac{\lambda_1}{\lambda_1-\lambda_2}\right)e^{\lambda_2 t}\right] \quad ...(14)$$

$$= y_o\frac{\lambda_1\lambda_2}{\lambda_2-\lambda_1}\left[\frac{1}{\lambda_1}e^{\lambda_1 t} - \frac{1}{\lambda_2}e^{\lambda_2 t}\right] \quad ...(15)$$

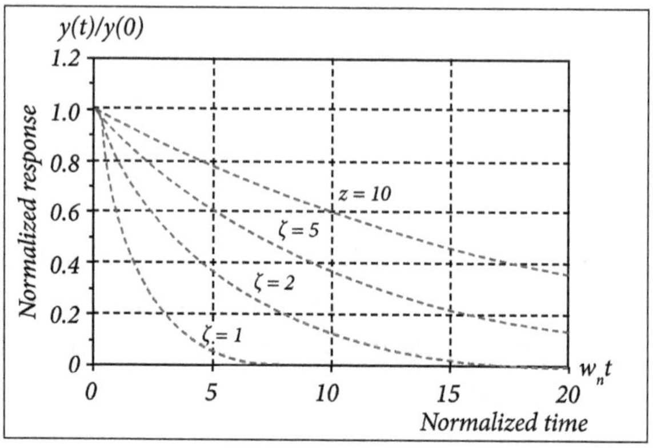

Homogeneous response of an over damped and critically damped second-order system for the initial condition $y(0)=y_o$ and $\dot{y}(0)=0$

If the roots of the characteristic equation are identical $\lambda_1 = \lambda_2 = \lambda$, the solution is based on equation (9) and is,

$$y_h(t) = y_o\left[e^{\lambda t} - \lambda t e^{\lambda t}\right] \quad ...(16)$$

The system response depends directly on the values of the damping ratio ζ and the undamped natural frequency ω_n.

Over-damped System ($\zeta > 1$)

When the damping ratio ζ is greater than one, the two roots of the characteristic equation are real and negative,

$$\lambda_1, \lambda_2 = \omega_n\left(-\zeta \pm \sqrt{\zeta^2 - 1}\right) \quad ...(17)$$

From equation (15), the homogeneous response is obtained as,

$$y_h(t) = y_o\left[\frac{-\zeta + \sqrt{\zeta^2-1}}{2\sqrt{\zeta^2-1}}e^{\left(-\zeta-\sqrt{\zeta^2-1}\right)\omega_n t} - \frac{-\zeta - \sqrt{\zeta^2-1}}{2\sqrt{\zeta^2-1}}^{\left(-\zeta+\sqrt{\zeta^2-1}\right)\omega_n t}\right] \quad ...(18)$$

Which is the sum of two decaying real exponentials, each with a different decay rate that defines a time constant,

$$\tau_1 = -\frac{1}{\lambda_1}, \qquad \tau_2 = -\frac{1}{\lambda_2} \qquad \text{...(19)}$$

The response exhibits no overshoot or oscillation and is known as an over damped response. The above figure shows this response as the function of ζ using the normalized time scale of $\omega_n t$.

Critically Damped System ($\zeta = 1$)

When the damping ratio $\zeta = 1$, the roots of the characteristic equation are real and identical,

$$\lambda_1 = \lambda_2 = -\omega_n \qquad \text{...(20)}$$

The solution to the initial condition response is found from equation (16),

$$yh(t) = y_o\left[e^{-\omega_n t} + \omega_n t e^{-\omega_n t}\right] \qquad \text{...(21)}$$

This response form is known as a critically damped response because it marks the transition between the non-oscillatory over-damped response and the oscillatory response.

Under-damped System ($0 \le \zeta < 1$)

When the damping ratio is greater than or equal to zero but less than 1, the two roots of the characteristic equation are complex conjugates with negative real parts,

$$\lambda_1, \lambda_2 = -\zeta\omega_n \pm j\omega_n\sqrt{1-\zeta^2} = -\zeta\omega_n \pm j\omega_d \qquad \text{...(22)}$$

Where $j = \sqrt{1}$ and where ω_d is defined to be the damped natural frequency,

$$\omega_d = \omega_n\sqrt{1-\zeta^2} \qquad \text{...(23)}$$

The response may be determined by substituting the values of the roots in equation (22) into (15),

$$y_h(t) = y_o\left[\left(\frac{-\zeta\omega_n - j\omega d}{-2j\omega_d}\right)e^{(-\zeta\omega_n + j\omega_d)t} + \left(\frac{-\zeta\omega_n + j\omega_d}{2j\omega_d}\right)e^{(-\zeta\omega_n - j\omega_d)t}\right]$$

$$= y_o e^{-\zeta\omega_n t}\left[\frac{e^{+j\omega_d t} + e^{-j\omega_d t}}{2} + \left(\frac{\zeta\omega_n}{\omega_d}\right)\frac{e^{j\omega_d t} - e^{-j\omega_d t}}{2j}\right] \qquad \text{...(24)}$$

When the Euler identities $\cos\alpha = (e^{+j\alpha} + e^{-j\alpha})/2$ and $\sin\alpha = (e^{+j\alpha} - e^{-j\alpha})/2j$ are substituted in the solution is,

$$y_h(t)=y_o e^{-\zeta\omega_n t}\left[\cos\omega_d t+\frac{\zeta\omega n}{\omega_d}\sin\omega_d t\right]$$

$$=y_o \frac{e^{-\zeta\omega_n t}}{\sqrt{1-\zeta^2}}\cos(\omega_d t-\psi) \qquad ...(25)$$

Where the phase angle ψ is,

$$\psi=\tan^{-1}\frac{\zeta}{\sqrt{1-\zeta^2}} \qquad ...(26)$$

For damping ratios near unity, the response decays rapidly with few oscillations, but as the damping is decreased and approaches zero, the response becomes increasingly oscillatory. When the damping is zero, the response becomes a pure oscillation and persists for all the time.

$$yh(t)=y_o\cos(\omega_n t) \qquad ...(27)$$

The decay ratio is unity if the damping ratio is zero and decreases as the damping ratio increases, reaching a value of zero as the damping ratio approaches unity.

Unstable System ($\zeta < 0$)

If the damping ratio is negative, the roots to the characteristic equation have positive real parts and the real exponential term grows in an unstable fashion. When $-1<\zeta<0$, the response is oscillatory with an overall exponential growth in amplitude while the solution for $\zeta<-1$ grows as a real exponential.

6.3 Problem Solving using R-L-C Elements with DC Excitation and AC Excitation

DC Response

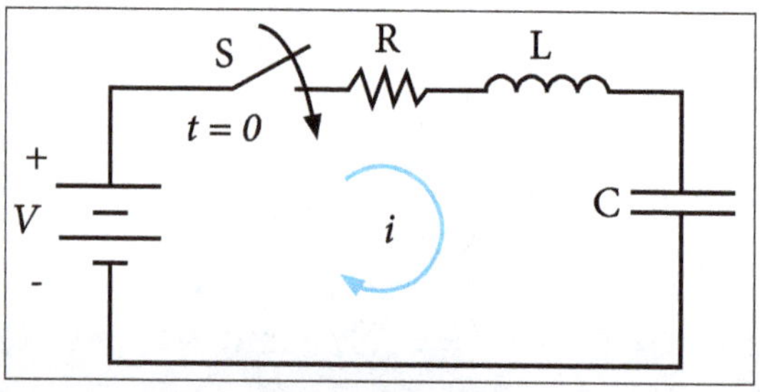

DC Response of RLC circuit.

By applying KVL,

$$V = Ri + L\frac{di}{dt} + \frac{1}{C}\int i\,dt$$

Differentiate on both sides, $0 = R\frac{di}{dt} + L\frac{d^2i}{dt^2} + \frac{1}{C}i$

Rearranging the above equation, $\frac{d^2i}{dt^2} + \frac{R}{L}\frac{di}{dt} + \frac{1}{LC}i + 0$

Above equation, can be written as, $D^2 + \frac{R}{L}D + \frac{1}{LC} = 0$

By solving the equation, $D_1, D_2 = -\frac{R}{2L} \pm \sqrt{\left(\frac{R}{2L}\right)^2 - \frac{1}{LC}}$

$$D_1, D_2 \Rightarrow K_1 \pm K_2$$

Let assume, $K_1 = -\frac{R}{2L}, K_2 = \sqrt{\left(\frac{R}{2L}\right)^2 - \frac{1}{LC}}$

Based on K_2 value, solution becomes:

Case (i)

If K_2 is positive, then the roots become real and different gives over damped response,

$$i(t) = c_1 e^{[K_1 + K_2]t} + c_2 e^{[K_1 - K_2]t}$$

Case (ii)

If K_2 is zero, then the roots become real and same gives critically damped response,

$$i(t) = e^{K_1 t}[c_1 t + c_2]$$

Case (iii)

If K_2 is negative, then the roots become complex conjugate gives under-damped response,

$$i(t) = e^{K_1 t}[c_1 \cos K_2 t + c_2 \sin K_2 t]$$

Time Response

Case (i)

If K_2 is positive, then the roots become real and different gives over damped response,

$$i(t)=c_1e^{[K_1+K_2]t}+c_2e^{[K_1-K_2]t}$$

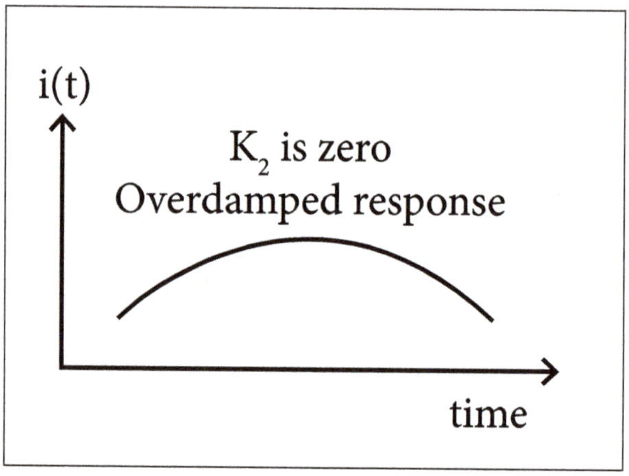

Case (ii)

If K_2 is zero, then the roots become real and same gives critically damped response,

$$i(t)=e^{K_1t}[c_1t+c_2]$$

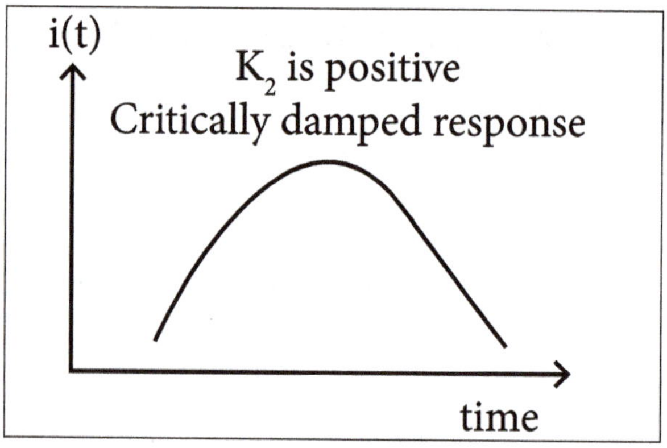

Case (iii)

If K_2 is negative, then the roots become complex conjugate gives under-damped response.

$$i(t)=e^{K_1t}[c_1\cos K_2t+c_2\sin K_2t]$$

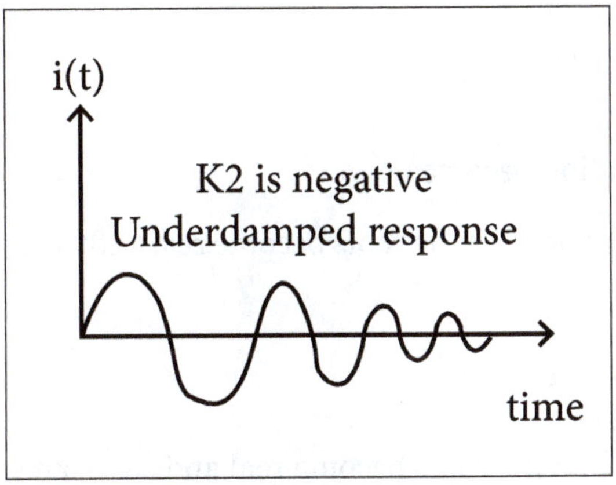

Ac (or) Sinusoidal Response of RLC Circuit

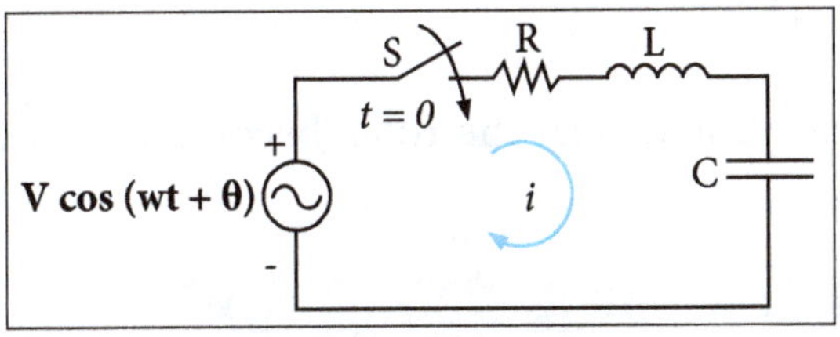

RLC circuit.

Step (1) by applying KVL,

$$Ri(t)+L\frac{di(t)}{dt}+\frac{1}{C}\int i\,(t)dt=V\cos(wt+\theta)$$

Differentiate on both sides,

$$R\frac{di(t)}{dt}+L\frac{d^2i(t)}{dt^2}+\frac{1}{C}i(t)=-Vw\sin(wt+\theta)$$

Divide both sides by L, then,

$$\frac{d^2i(t)}{dt^2}+\frac{R}{L}\frac{di(t)}{dt}+\frac{1}{LC}i(t)=-\frac{Vw}{L}\sin(wt+\theta) \quad ...(1)$$

The complete solution,

$$i(t)=i_c+i_p \quad ...(2)$$

Step (2) to find i_c, equate RHS of equation (1) = 0,

$$\frac{d^2i(t)}{dt^2}+\frac{R}{L}\frac{di(t)}{dt}+\frac{1}{LC}i(t)=0$$

$$\left(D^2+\frac{R}{L}D+\frac{1}{LC}\right)i(t)=0$$

Based on K_2 value, solution becomes,

Case (i) If K_2 is positive, and then the roots become real and different gives over damped response.

$$i(t)=c_1e^{[K_1+K_2]t}+c_2e^{[K_1-K_2]t}$$

Case (ii) If K_2 is zero, then the roots become real and same gives critically damped response.

$$i(t)=e^{K_1t}[c_1t+c_2]$$

Case (iii) If K_2 is negative, and then the roots become complex conjugate gives under-damped response.

$$i(t)=e^{K_1t}[c_1\cos K_2t+c_2\sin K_2t]$$

Step (3) to find the particular integral,

$$i_p=A\cos(wt+\theta)+B\sin(wt+\theta)$$

$$\frac{di_p}{dt}(or)i_p'=-Aw\sin(wt+\theta)+Bw\cos(wt+\theta)$$

$$\frac{d^2i}{dt^2}=-Aw^2\cos(wt+\theta)-Bw^2\sin(wt+\theta)$$

Substitute i_p, i_p' and i_p" in equation (1),

Then,

$$-Aw^2\cos(wt+\theta)-Bw^2\sin(wt+\theta)+$$

$$\frac{R}{L}[-Aw\sin(wt+\theta)+Bw\cos(wt+\theta)]$$

$$\frac{1}{RC}[A\cos(wt+\theta)+B\sin(wt+\theta)]=-\frac{Vw}{L}\sin(wt+\theta)$$

By equating cos and sin term on both sides,

$$\left.\begin{aligned} -Aw^2 + Bw\frac{R}{L} + A\frac{1}{RC} &= 0 \\ -Bw^2 - A\frac{R}{L} + B\frac{1}{RC} &= -\frac{Vw}{L} \end{aligned}\right\} \text{by solving this,}$$

$$A = \frac{VR}{R^2 + \left(\frac{1}{wC} - wL\right)^2}; \quad B = \frac{-V\left(\frac{1}{wC} - wL\right)}{R^2 + \left(\frac{1}{wC} - wL\right)^2}$$

Let us consider $A = M\cos\phi$, $B = M\sin\phi$, then subs in equation i_p,

$$i_p = M\cos\Phi\cos(wt+\theta) + M\sin\Phi\sin(wt+\theta)$$

Rewriting the above equation,

$$i_p = M\cos(wt+\theta)\cos\Phi + M\sin(wt+\theta)\sin\Phi$$

$$[\text{wkt } \cos A\cos B + \sin A\sin B = \cos(A-B)]$$

$$i_p = M\cos(wt+\theta-\Phi) \qquad \ldots(3)$$

By solving $A = M\cos\Phi$, $B = M\sin\Phi$;

$$M = \frac{V}{\sqrt{R^2 + \left(\frac{1}{wC} - wL\right)^2}}; \Phi = -\tan^{-1}\left(\frac{1}{wCR} - \frac{wL}{R}\right)$$

Step (4) by substituting M and ϕ in equation (3),

$$i_p = \frac{V}{\sqrt{R^2 + \left(\frac{1}{wC} - wL\right)^2}}\cos\left[wt + \theta + \tan^{-1}\left(\frac{1}{wCR} - \frac{wL}{R}\right)\right]$$

Step (5) the complete solution $i(t) = i_c + i_p$.

6.4 Response Related to s-plane Rotation of Roots and Solutions using Laplace Transform Method

A transform is a change in the mathematical description of a physical variable to facilitate,

$$\mathscr{L}\{f(t)\}=F(s)=\int_0^\infty f(t)e^{-st}\,dt$$

Here the complex frequency is $s=\sigma+j\omega$. Since the argument of the exponent e in the above equation must be dimensionless, it follows that has the dimensions of frequency and units of inverse seconds (sec^{-1}).

The notation implies that once the integral has been evaluated, f(t), a time domain function is transformed to F(s), a frequency domain function. If the lower limit of integration is $-\infty$, then it is called the bilateral Laplace transform.

However for circuit applications, the lower limit is taken as zero and accordingly the transform is unilateral in nature. The lower limit of integration is sometimes chosen to be -0 to permit f(t) to include δ(t) or its derivatives. Thus, we should note immediately that the integration from -0 to +0 is zero except when an impulse function or its derivatives are present at the origin.

Region of Convergence

The Laplace transform of a signal f(t) is an integral operation. It exists if $f(t)e^{-\sigma}$ is absolutely integrable. That is $\int_0^\infty f(t)e^{-\sigma t}\,dt<\infty$ only typical choices of σ will make the integral converge. The range of σ that ensures the existence of X(s) defines the region of convergence (ROC) of the Laplace transform.

$$X=(s)=\int_0^\infty x(t)e^{-(\sigma+j\omega)t}dt$$

Functional Transform

A functional transform is simply the Laplace transform of a specified function of t. Here we assume that f(t) is zero for t < 0.

Decaying Exponential Function

$$f(t)=e^{-at}u(t)$$

Where,

$a > 0$ and u(t) is the unit step function,

$$L = \{e^{-at}u(t)\} = F(s) = \int_0^\infty f(t)dt$$

$$= \int_0^\infty e^{-at}e^{-st}dt$$

$$= \left.\frac{-e^{-(s+a)t}}{(s+a)}\right|_0^\infty$$

$$= \frac{1}{s+a}$$

Unit Step Function

$$f(t) = u(t)$$

$$L\{u(t)\} = F(s) = \int_0^\infty e^{-st}dt = \frac{1}{s}$$

Impulse Function

$$f(t) = \delta(t)$$

$$L\{\delta(t)\} = F(s) = \int_{0^-}^\infty \delta(t)e^{-st}dt = e^{-st}\Big|_{t=0} = 1$$

Sinusoidal Function

$$f(t) = \sin\omega t, \quad t \geq 0$$

Since,

$$\sin\omega t = \frac{1}{2j}\left[e^{j\omega t} - e^{-j\omega t}\right]$$

And,

$$L\{e^{-at}\} = \frac{1}{s+a}$$

We have,

$$L\ \{\sin \omega t\} = F(s) = \frac{1}{2j}\int_{0}^{\infty}\left(e^{j\omega t} - e^{-j\omega t}\right)e^{-st}dt$$

$$= \frac{1}{2j}\left[\frac{1}{s - j\omega} - \frac{1}{s + j\omega}\right]$$

$$= \frac{\omega}{s^2 + \omega^2}$$

List of Important Laplace Transform Pairs

$f(t)(t \geq 0)$	$F(s)$
$\delta(t)$	1
$u(t)$	$\frac{1}{s}$
t	$\frac{1}{s^2}$
e^{-at}	$\frac{1}{s+a}$
$\sin \omega t$	$\frac{\omega}{s^2 + \omega^2}$
$\cos \omega t$	$\frac{s}{s^2 + \omega^2}$
t^n	$\frac{n!}{s^{n+1}}$
te^{-at}	$\frac{1}{(s+a)^2}$
$e^{-at}\sin \omega t$	$\frac{\omega}{(s+a)^2 + \omega^2}$
$e^{-at}\cos \omega t$	$\frac{s+a}{(s+a)^2 + \omega^2}$

This includes the functions of most interest in an introductory course on circuit applications.

Properties of Laplace Transform

	Property/ Theorem	Time Domain	Complex Frequency Domain
1.	Linearity	$c_1f_1(t)+c_2f_2(t)+...+c_nf_n(t)$	$c_1F_1(s)+c_2F_2(s)+...+c_nF_n(s)$
2.	Time Shifting	$f(t-a)u_o(t-a)$	$e^{-as}F(s)$
3.	Frequency Shifting	$e^{-as}f(t)$	$F(s+a)$
4.	Time Scaling	$f(at)$	$\frac{1}{a}F\left(\frac{s}{a}\right)$
5.	Time Differentiation	$\frac{d}{dt}f(t)$	$sF(s)+f(o^-)$
6.	Frequency Differentiation	$tf(t)$	$-\frac{d}{ds}F(s)$
7.	Time Integration	$\int_{-\infty}^{t} f(\tau)d\tau$	$\frac{F(s)}{s}+\frac{f(o^-)}{s}$
8.	Frequency Integration	$\frac{f(t)}{t}$	$\int_s^{\infty} F(s)ds$
9.	Time Periodicity	$f(t+nT)$	$\frac{\int_o^T f(t)e^{-st}dt}{1-e^{-sT}}$
10.	Initial Value Theorem	$\lim_{t\to 0} f(t)$	$\lim_{s\to\infty} sF(s)=f(o^-)$
11.	Final Value Theorem	$\lim_{t\to\infty} f(t)$	$\lim_{s\to o} sF(s)=f(\infty)$ $s\to 0$
12.	Time Convolution	$f_1(t)=*f_2(t)$	$F_1(s)F_2(s)$
13.	Frequency Convolution	$f_1(t)\,f_2(t)$	$\frac{1}{2\pi j}F_1(s)*F_2(s)$

RC, RL Circuits using Laplace Transform for DC Input

DC Response of RL circuit with initial i(o) = o	Time Response of RL circuit with DC input
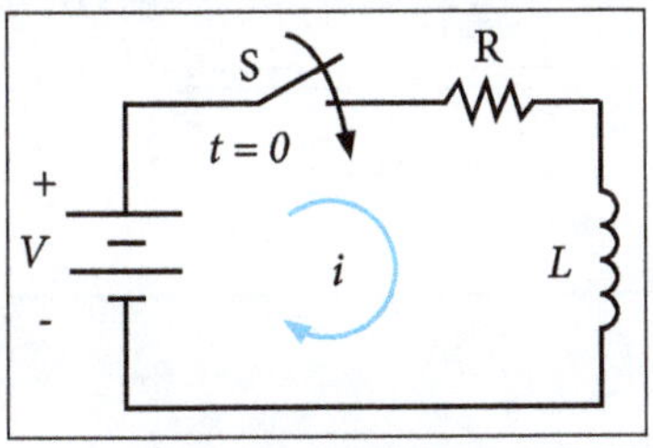	The current equation for RL circuit is $i(t)=\frac{V}{R}\left(1-e^{-Rt/L}\right)$

By applying KVL,

$$V=Ri(t)+L\frac{di}{dt}$$

Take Laplace Transform on both sides

$$\frac{V}{s}=RI(s)+L\left[sI(s)-i(0)\right]$$

$$\frac{V}{s}=\left[R+sL\right]I(s)$$

$$I(s)=\frac{V}{s\left[sL+R\right]}=\frac{V}{L}*\frac{1}{s(s+sR/L)}$$

By using laplace transform

$$\frac{1}{s(s+R/L)}=\frac{A}{s}+\frac{B}{s+R/L}$$

$$1=A(s+R/L)+Bs$$

$$\text{sub } s=0,1=A(R/L)\Rightarrow A=\frac{L}{R}$$

$$\text{sub} ss=-\frac{R}{L},1=B(-R/L)\Rightarrow B=-\frac{L}{R}$$

$$I(s)=\frac{V}{L}*\left[\frac{L/R}{s}-\frac{L/R}{s+\frac{R}{L}}\right]$$

$$I(s)=\frac{V}{R}\left[\frac{1}{s}-\frac{1}{s+R/L}\right]$$

Take Laplace inverse transform, then

$$i(t)=\frac{V}{R}\left(1-e^{-Rt/L}\right)$$

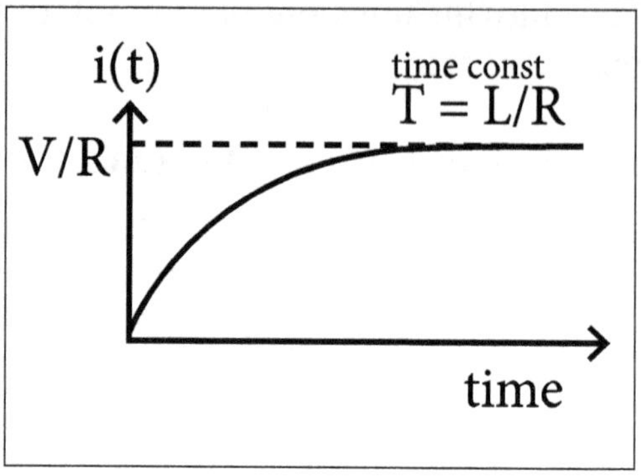

(a)Voltage across Resistor

$$V_R=R*i=R\frac{V}{R}\left[1-e^{-\frac{Rt}{L}}\right]$$

$$V_R=V\left[1-e^{-\frac{Rt}{L}}\right]$$

(b) Voltage across Inductor

$$V_L=L\frac{di}{dt}$$

$$V_L=L\frac{d}{dt}\left(\frac{V}{R}\left[1-e^{-\frac{Rt}{L}}\right]\right)$$

$$V_L=L\frac{V}{R}\left[-e^{-\frac{Rt}{L}}\right]*\left(\frac{-R}{L}\right)=Ve^{-\frac{Rt}{L}}$$

Time response of V_R And VL

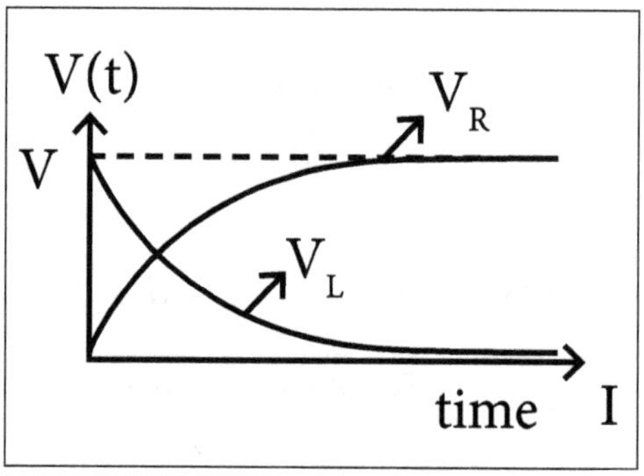

Answer: $i(t)=\frac{V}{R}\left(1-e^{-Rt/L}\right)$

DC Response of RC circuit with initial voltage across capacitor v(0) = 0

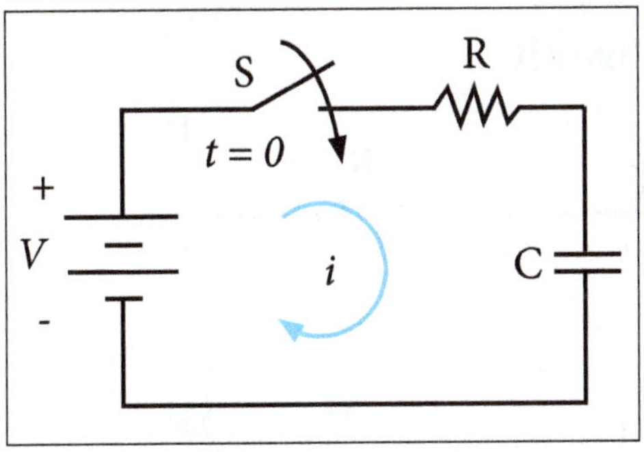

By applying KVL,

$$V = Ri(t) + \frac{1}{C}\int i\,dt$$

Take Laplace Transform on both side

$$\frac{V}{s} = RI(s) + \frac{1}{Cs}I(s)$$

$$\frac{V}{s} = \left[R + \frac{1}{Cs}\right]I(s)$$

$$\frac{V}{s} = \left[\frac{RCs+1}{Cs}\right]I(s)$$

$$I(s) = \frac{VC}{[RCs+1]}$$

Take RC Commonly out from Denominator r

$$I(s) = \frac{VC}{RC} * \frac{1}{(s+1/RC)}$$

$$I(s) = \frac{V}{R} * \frac{1}{(s+1*/RC)}$$

By using Laplace inverse transform

$$i(t) = \frac{V}{R} * e^{-t/RC}$$

Answer: $i(t) = \frac{V}{R} * e^{-t/RC}$

Time Response of RC circuit with DC input

The current equation for RC circuit is

$$i(t) = \frac{V}{R}e^{-t/RC}$$

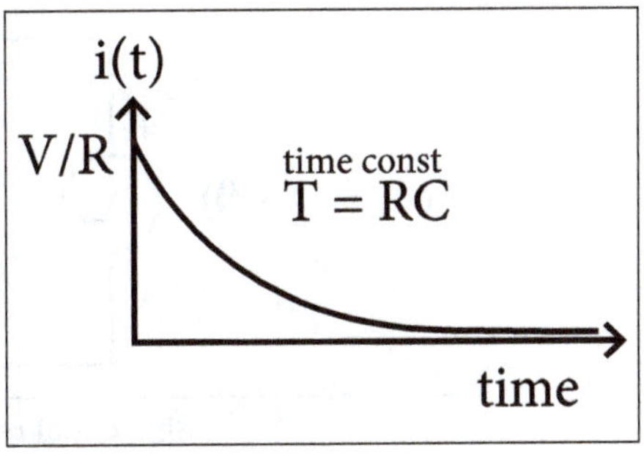

(a) Voltage across Resistor

$$V_R = R^* i(t) = R\frac{V}{R}e^{\frac{-t}{RC}}$$

$$V_R = Ve^{\frac{-t}{RC}}$$

(b) Voltage across capacitor

$$V_C = \frac{1}{C}\int_0^t i\,dt$$

$$V_C = \frac{1}{C}\int_0^t \frac{V}{R}e^{-\frac{t}{RC}}dt$$

$$V_C = \frac{V}{RC}\left[-RCe^{-\frac{t}{RC}}\right]_0^t = V\left[1 - e^{-\frac{t}{RC}}\right]$$

Time Response of $V_{R \text{ and } VL}$

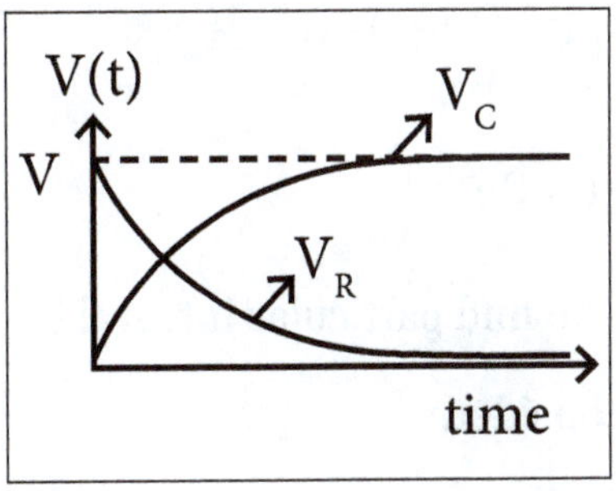

6.4.1 RC, RL Circuits using Laplace Transform for AC or Sinusoidal Input

1. AC or Sinusoidal Response of RL Circuit

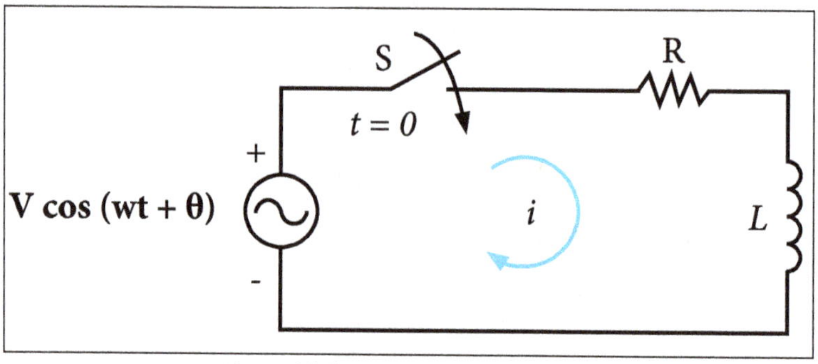

Sinusoidal response of RL circuit.

Step (1) by applying KVL,

$$Ri(t)+L\frac{di(t)}{dt}=V\cos(wt+\theta)$$

Divide both sides by L, then,

$$\frac{di(t)}{dt}+\frac{R}{L}i(t)=\frac{V}{L}\cos(wt+\theta) \qquad ...(1)$$

The complete solution $i(t)=i_c + i_p$ ...(2)

Step (2) to find ic, equate RHS of equation (1) = 0,

$$\frac{di(t)}{dt}+\frac{R}{L}i(t)=0$$

$$\left(D+\frac{R}{L}\right)i(t)=0$$

$$i_c=Ce^{-Rt/L}$$

Step (3) to find particular integral i_p,

Let us consider,

$$i_p=A\cos(wt+\theta)+B\sin(wt+\theta)$$

$$\frac{di_p}{dt} \text{ or } i_p'=-Aw\sin(wt+\theta)Bw\cos(wt+\theta)$$

Substitute i_p and i_p' in equation (1) then,

$$-Aw\sin(wt+\theta)Bw\cos(wt+\theta)+\frac{R}{L}A\cos(wt+\theta)$$

$$+\frac{R}{L}B\sin(wt+\theta)=\frac{V}{L}\cos(wt+\theta)$$

Equate cos and sin term on both sides,

$$\left.\begin{array}{l} A\frac{R}{L}+Bw=\frac{V}{L} \\ -Aw+B\frac{R}{L}=0 \end{array}\right\}\text{by solving this,}$$

$$A=\frac{VR}{R^2+(wL)^2}$$

$$B=\frac{V(wL)}{R^2(wL)^2}$$

Let us consider $A = M\cos\phi$, $B = M\sin\phi$, then subs in equation i_p,

$$i_p=M\cos\Phi\cos(wt+\theta)+M\sin\Phi\sin(wt+\theta)$$

Rewriting the above equation,

$$i_p=M\cos(wt+\theta)\cos\Phi+M\sin(wt+\theta)\sin\Phi$$

$$\left[\text{wkt } \cos A\cos B + \sin A\sin B = \cos(A-B)\right]$$

$$i_p=M\cos(wt+\theta-\Phi) \qquad ...(3)$$

By solving $A=M\cos\Phi$, $B = M\sin\Phi$;

$$M^2\cos^2\Phi+M^2\sin^2\Phi=A^2+B^2$$

$$M^2\left[\cos^2\Phi+\sin^2\Phi\right]=\frac{V^2R^2}{\left(R^2+(wL)^2\right)^2}+\frac{V^2(wL)^2}{\left(R^2+(wL)^2\right)^2}$$

$$M^2 = \frac{V^2R^2 + (wL)^2}{\left(R^2 + (wL)^2\right)^2}$$

$$M = \frac{V}{\sqrt{\left(R^2 + (wL)^2\right)}}$$

$$\frac{M \sin\Phi}{M \cos\Phi} = \frac{B}{A} \Rightarrow \tan\Phi = \left(\frac{wL}{R}\right)$$

$$\Phi = \tan^{-1}\left(\frac{wL}{R}\right)$$

Step (4) by substituting M and ϕ in equation (3),

$$i_p = \frac{V}{\sqrt{R^2 + (wL)^2}} \cos\left[wt + \theta - \tan^{-1}\left(\frac{wL}{R}\right)\right]$$

Step (5) the complete solution becomes,

$$i(t) = i_c + i_p$$

$$i(t) = ce^{-\frac{Rt}{L}} + \frac{V}{\sqrt{R^2 + (wL)^2}} \cos\left[wt + \theta - \tan^{-1}\left(\frac{wL}{R}\right)\right]$$

2. AC or Sinusoidal Response of RC Circuit

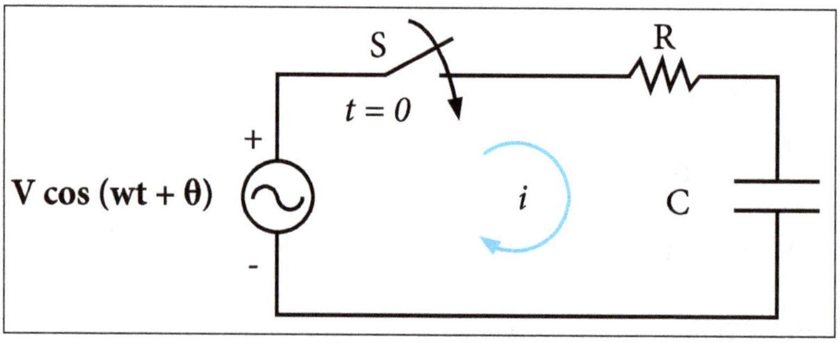

Sinusoidal response of RC circuit.

Step (1) by applying KVL,

$$Ri(t) + \frac{1}{C}\int i(t)dt = V\cos(wt + \theta)$$

Differentiate on both sides,

$$R\frac{di(t)}{dt}+\frac{1}{C}i(t)=-Vw\sin(wt+\theta)$$

Divide both sides by R, then,

$$\frac{di(t)}{dt}+\frac{1}{RC}i(t)=-\frac{Vw}{R}\sin(wt+\theta)$$

The complete solution $i(t)=i_c+i_p$ ...(2)

Step (2) to find ic, equate RHS of equation (1) = 0,

$$\frac{di(t)}{dt}+\frac{1}{RC}i(t)=0$$

$$\left(D+\frac{1}{RC}\right)i(t)=0$$

$$i_c=Ce^{-t/RC}$$

Step (3) to find particular integral i_p,

Let us consider,

$$i_p=A\cos(wt+\theta)+B\sin(wt+\theta)$$

$$\frac{di_p}{dt}(\text{or})i_p'=-Aw\sin(wt+\theta)+Bw\cos(wt+\theta)$$

Substitute i_p and i_p' in equation (1) then,

$$-Aw\sin(wt+\theta)+Bw\cos(wt+\theta)+\frac{1}{RC}A\cos(wt+\theta)$$

$$+\frac{1}{RC}B\sin(wt+\theta)=-\frac{Vw}{R}\sin(wt+\theta)$$

Equate cos and sin term on both sides,

$$\left.\begin{aligned}A\frac{1}{RC}+Bw&=0\\-Aw+B\frac{1}{RC}&=-\frac{Vw}{R}\end{aligned}\right\}\text{by solving this,}$$

$$A = \frac{VR}{R^2 + (1/wC)^2}$$

$$B = \frac{-V(1/wC)^2}{R^2 + (1/wC)^2}$$

Let us consider $A = M\cos\phi$, $B = M\sin\phi$, then subs in equation i_p,

$$i_p = M\cos\Phi\cos(wt+\theta) + M\sin\Phi\sin(wt+\theta)$$

Rewriting the above equation,

$$i_p = M\cos(wt+\theta)\cos\Phi + M\sin(wt+\theta)\sin\Phi$$

[we know that, cosAcosB + sinAsinB = cos(A – B)]

$$i_p = M\cos(wt+\theta-\Phi) \quad ...(3)$$

By solving $A = M\cos\Phi$, $B = M\sin\Phi$;

$$M^2\cos^2\Phi + M^2\sin\Phi = A^2 + B^2$$

$$M^2\left[\cos^2\Phi + \sin^2\Phi\right] = \frac{V^2R^2}{\left(R^2(1/wC)^2\right)^2} + \frac{V^2(1/wC)^2}{\left(R^2(1/wL)^2\right)^2}$$

$$M^2 = \frac{V^2R^2 + (1/wC)^2}{\left(R^2(1/wC)^2\right)^2}$$

$$M = \frac{V}{\sqrt{R^2(1/wC)^2}}$$

$$\frac{M\sin\Phi}{M\cos\Phi} = \frac{B}{A} \Rightarrow \tan\Phi = -\left(\frac{1/wC}{R}\right)$$

$$\Phi = -\tan^{-1}\left(\frac{1}{wCR}\right)$$

Step (4) By substituting M and ϕ in equation (3),

$$i_p = \frac{V}{\sqrt{R^2 (1/wC)^2}} \cos\left[wt + \theta + \tan^{-1}\left(\frac{1}{wCR}\right)\right]$$

Step (5) The complete solution becomes,

$$i(t) = i_c + i_p$$

$$i(t) = ce^{-t/RC} + \frac{V}{\sqrt{R^2 (1/wC)^2}} \cos\left[wt + \theta + \tan^{-1}\left(\frac{1}{wCR}\right)\right].$$

Permissions

All chapters in this book are published with permission under the Creative Commons Attribution Share Alike License or equivalent. Every chapter published in this book has been scrutinized by our experts. Their significance has been extensively debated. The topics covered herein carry significant information for a comprehensive understanding. They may even be implemented as practical applications or may be referred to as a beginning point for further studies.

We would like to thank the editorial team for lending their expertise to make the book truly unique. They have played a crucial role in the development of this book. Without their invaluable contributions this book wouldn't have been possible. They have made vital efforts to compile up to date information on the varied aspects of this subject to make this book a valuable addition to the collection of many professionals and students.

This book was conceptualized with the vision of imparting up-to-date and integrated information in this field. To ensure the same, a matchless editorial board was set up. Every individual on the board went through rigorous rounds of assessment to prove their worth. After which they invested a large part of their time researching and compiling the most relevant data for our readers.

The editorial board has been involved in producing this book since its inception. They have spent rigorous hours researching and exploring the diverse topics which have resulted in the successful publishing of this book. They have passed on their knowledge of decades through this book. To expedite this challenging task, the publisher supported the team at every step. A small team of assistant editors was also appointed to further simplify the editing procedure and attain best results for the readers.

Apart from the editorial board, the designing team has also invested a significant amount of their time in understanding the subject and creating the most relevant covers. They scrutinized every image to scout for the most suitable representation of the subject and create an appropriate cover for the book.

The publishing team has been an ardent support to the editorial, designing and production team. Their endless efforts to recruit the best for this project, has resulted in the accomplishment of this book. They are a veteran in the field of academics and their pool of knowledge is as vast as their experience in printing. Their expertise and guidance has proved useful at every step. Their uncompromising quality standards have made this book an exceptional effort. Their encouragement from time to time has been an inspiration for everyone.

The publisher and the editorial board hope that this book will prove to be a valuable piece of knowledge for students, practitioners and scholars across the globe.

Index

www.ingramcontent.com/pod-product-compliance
Lightning Source LLC
Chambersburg PA
CBHW061939190326
41458CB00009B/2773
9781647285074